The Metamorphosis of War

At the Interface

Volume 80
A volume in the *Probing the Boundaries* series
'War, Civil Conflict, Peace and Security'

Probing the Boundaries

The Metamorphosis of War

Edited by
Avery Plaw

Rodopi

Amsterdam - New York, NY 2012

The paper on which this book is printed meets the requirements of "ISO 9706:1994, Information and documentation - Paper for documents - Requirements for permanence".

ISBN: 978-90-420-3571-3
E-Book ISBN: 978-94-012-0846-8

Printed in the Netherlands

Table of Contents

Acknowledgements

Every book is a collaborative project, but an interdisciplinary collection of papers is especially so. Such a project is unlikely to reach completion without the patient and cooperative work of many. So it is hardly surprising that there are numerous people who deserve acknowledgement and gratitude for their shares in this book. The first of these is Axel Augé to whom I'm thankful for his critical contributions to the introduction. Aaron Cardoso, Kristen Emmett and Ryan Carreiro also have my deep gratitude for their invaluable help with editing and formatting the manuscript. I'd also like to express profound thanks to the Steering Group of the War, Virtual War and Human Security Project which organized the conference that brought these authors together and which authorized the publication of this book. Special thanks go to Robert Fisher and Graeme Goldsworthy, who led the conference, as well as to Bob Brecher, Andrew Wilson, Jones Irwin and Julia Boll for their work on it and the other publications that have resulted. A very special vote of thanks must also go to Anthony Garro, the Provost of the University of Massachusetts at Dartmouth, whose generous Publication Subvention Grant aided immeasurably in the preparation of this book. Last but not least, I would like to thank all the contributors who have literally made this book by sharing their reflections, analysis and insight.

Introduction: The Transformations of War

Avery Plaw and Axel Augé

This book is a product of the Fifth Global Conference on War, Virtual War and Human Security (WVW5) held in Budapest, Hungary from the fifth to the seventh of May 2008. The conference served as an interdisciplinary forum in which scholars attempted to come to grips with the rapidly changing character of modern armed conflict and contemporary responses to it. The following volume encompasses the most valuable contributions to that project. This introduction frames the chapters that follow by briefly sketching some important developments in the concepts, practices and purposes of war since the beginning of the nineteenth century.

1. The New Wars

Some eight years ago, Herfried Münkler pointed out how jarringly different 'new wars' look in comparison with 'classical inter-state wars' exemplified in the 'great battle' in which uniformed armies clash across a clear if shifting front.[1] Indeed, as he noted, the very language of war has been transformed in recent decades,[2] as formulas like asymmetrical, dissymmetrical, informal and unconventional war, low-intensity conflict and counterinsurgency warfare have risen to prominence.[3] Likewise, attention has shifted to unconventional participants in warfare like child soldiers, warlords, female suicide-bombers, paramilitary groups, death squads, guerrilla forces, private military contractors, and Predators, Reapers and other unmanned aerial vehicles. The rapid mutation and diversification of war has even called into question the continuing relevance of the conventional laws and norms that have traditionally regulated armed conflict.[4]

How then can we understand the increasingly diverse forms of violence that have come to share the nomenclature of war? How should we respond to them? The common premise of WVW5 and this book is that the best hope of answering these important questions lies in mobilizing and coordinating the full range of disciplines concerned with war and its effects in an interdisciplinary dialogue devoted to analysing these questions critically and synthesizing the results. The WVW5 conference actively promoted cross-fertilization between research in fields as diverse as strategic studies, literary theory, psychology, international relations, philosophy, communications and media studies, history and politics. This volume offers expanded versions of twelve chapters selected for publication. The final versions of these chapters reflect the productive exchange of ideas achieved through the conference. By consequence, these contributions reflect an active dialogue rather than a series of disparate reflections on related topics.

2. New Wars for Old

In the early nineteenth century the Prussian Major-General Carl von Clausewitz famously described war as 'policy by other means.'[5] He presented it as, in its essence, 'nothing but a duel on a larger scale,' or a wrestling match.[6] The figurative duellers or wrestlers, in his day, embodied states.[7] The object of the exercise, he believed, was to render the opponent 'incapable of further resistance' through the use of force, and thus to 'compel our enemy to do our will.'[8] In essence then, for Clausewitz, modern war was simply the means by which one state (or states) forcibly imposed its policy on another (or others).

Clausewitz and his work remain canonical for both scholars and practitioners of war today. Scholars, scientists and soldiers still continually invoke his famous dictum that war is 'merely the continuation of policy by other means.' What is often overlooked is the detailed attention that Clausewitz devoted to the ways in which the aims, means, tactics and strategies of war have metamorphosed over time. In Book Eight of *On War*, for example, he discusses 'the transformations of war':

> The semibarbarous Tartars, the republic of antiquity, the feudal lords and trading cities of the Middle Ages, eighteenth-century kings and the rulers and peoples of the nineteenth century—all conducted war in their own particular way, using different methods and pursuing different aims.[9]

Clausewitz goes on to detail the form of war that dominated each epoch, each with its distinctive aims, and argues that an acute awareness of the on-going transformations of war is essential to understanding its character in one's own day, and in particular to forming a successful plan of war. He attributes the defeat of Napoleon's opponents in 1805, 1806 and 1809, for example, to their failure to fully grasp the metamorphosis of war in their own time.[10] There is a deep irony then in the continued reliance of strategists and practitioners of war today on von Clausewitz's insightful examination of the form of war dominant at the beginning of the nineteenth century. He himself warns us of the dangers of such reliance.

Martin Van Creveld has forcefully argued that the new wars seen today depart radically from the Clausewitzian model.[11] He argues that, in contrast to nineteenth century inter-state wars decided by great battles, '[t]he great majority of wars since 1945 have been Low Intensity Conflicts. In terms of both casualties suffered and political results achieved, these wars have been incomparably more important than others,' particularly those few that have had a more classical Clausewitzian character (e.g., the Falklands, the First Gulf War or even the Iran-Iraq war).[12]

According to Van Creveld, these low intensity conflicts (LICs) 'unfold in 'less developed' parts of the world,' 'rarely...involve regular armies' and don't 'rely primarily on... high-technology collective weapons.'[13] They are also 'far more bloody than any other kind of war fought since 1945,' particularly in terms of civilian lives.[14] Indeed, he stresses that 'all over the world today, the traditional distinction between peoples and armies is being broken down by... Low-Intensity Conflicts.'[15]

Münkler too stresses the distinctive character of the 'new wars' over the last few decades. He points, in particular, to the collapse of 'the state's monopoly of war' and to its progressive privatization, along with the characteristic 'asymmetry of military force' in contemporary conflicts, and 'the successive autonomization of forms of violence.'[16] One consequence of these developments is the frequently protracted duration of contemporary armed conflict—'new wars may stretch over decades, with no end in sight.'[17]

After affirming the growing importance of 'Large-Scale civil wars, often aided and abetted by great powers' and the corresponding destruction of 'increasing numbers of non-combatants,' Charles Tilly identifies an important additional feature of the new wars.[18] He stresses that states have begun 'to displace or even expel populations as never before,' with the result that such 'wars generate refugees at a huge rate.'[19]

Van Creveld, Münkler and Tilly's analyses each then draw attention to some distinctive facets of the new wars, but each also draws attention to the same core facts: wars are no longer generally waged between traditional actors (states) for traditional political reasons (e.g., territorial conquest, ideology). Rather, war has acquired new actors, new means and new purposes. At least four distinctive characteristics can be identified.

First, wars are preponderantly civil in the contemporary era rather than between states. Steven David reports that between 1995 and 2005, fully 95% of armed conflicts occurred within countries rather than between them.[20] Even some of the remaining international conflict during this period, notably the US-led 'War on Terror,' revolved around struggles between states and non-state groups (such as Al Qaeda) rather than around traditional conflicts between states.

Second, parties to these mainly civil wars typically exhibit non-traditional motivations and goals. Scholars like Mary Kaldor, Christopher Clapham and Colin McInnes have shed light on the political dimensions of these new wars.[21] They demonstrate that the motivations of combatants are typically cultural, religious or economic. They also emphasize that many combatants are irregular, that civilians have become the targets in combat, and that the use of force is directed to generating publicity as much as to achieving decisive physical victories.[22]

Third, more civilians die in the new wars than soldiers. Kaldor reports that at the beginning of the 20th century the ratio of military to civilian casualties in war was 8:1; in the wars of the 1990s the ratio was reversed, to one combatant for every eight civilian casualties.[23] One reason for this is that population centres often serve as new fields of battle.[24] Cities and towns like Bunia or Kinshasa (Democratic Republic of Congo), Abidjan (Ivory Coast), Mogadishu (Somalia), Grozny (Russia), Mitrovica (Kosovo) and Sarajevo (Bosnia and Herzegovina) function as the stage for violence in these new conflicts.

Fourth, full victory cannot be achieved by brute force alone. As Van Creveld, for example, convincingly shows, no major military power has been able to win a low intensity conflict (at least since World War II) by force of arms.[25] Even the vast military might of the two cold war superpowers were insufficient to assure victory against third world guerrilla forces that seized the political momentum in Vietnam and Afghanistan.[26] In modern conflicts, 'territory is captured,' in Mary Kaldor's apt phrase, 'through political control of the population rather than through military advance.'[27] Political strategy, control of information, rhetoric and propaganda filtered in part through the mass media, serve as crucial weapons in contemporary wars. Ultimately, the new wars are at least as much about winning converts as about clashing armed forces. Victory is impossible without the transformation of popular perceptions, in particular those of civilians, whether by conversion or coercion or some combination thereof. Consequently, the battles fought on the ground, and the blood spilled, are often strategically choreographed to engage and influence a popular audience through the emotions of anger, pity and fear. Acts of terrorism, with their focus on publicity and intimidation exemplify this pattern—indeed, Bruce Hoffman has aptly characterized them as 'a perverted form of show business.'[28] In a strange and terrible inversion, violence often becomes a mere publicity instrument rather than the axis of struggle: in an inversion of Clausewitz's famous dictum, it is politics that becomes war by other means.

Ultimately, the confused and grinding horror of the new wars may come to exceed the atrocities of the great wars of the first half of the twentieth century. The terrorist attacks in New York and elsewhere in the United States (11 September 2001), in Madrid (11 March 2004) in London (7 July 2005) and in Mumbai (26 to 29 November 2008) are indicators of the new wars' grim promise. The deliberate mass and indiscriminate killing of civilians in recent years makes it imperative to work collaboratively and across disciplinary and professional boundaries to comprehend the stakes, mechanisms and consequences of contemporary wars so that we may break the cycles of violence.

3. Chapter Overview

The Metamorphosis of War represents a step in this process. Its chapters seek to move us beyond the paradigms through which we have traditionally viewed war. They attempt to conceptualize war in ways that open paths toward new solutions. The book is divided into four sections, each addressing its own set of questions.

The first section focuses on the concept of war itself, how it is changing and how it needs to change further. Nick Mansfield begins the section by exposing the false duality of war and peace. He examines a range of ways that leading thinkers have distinguished the two. Specifically, he argues that there are at least three classical approaches. According to Hobbes war is natural and peace is an exceptional condition. For Kant, human beings are rational and peaceful by nature and fight only through ignorance and confusion. Clausewitz, by contrast, presents war as a mere extension of politics, so that policies of peace and war are equally natural and even linked. Mansfield concludes that none of the distinctions is satisfactory, and indeed that each reveals how war and peace infuse one another in their very definitions. He concludes that we need to understand not only the differences but also their mutual entanglement. In a similar vein, Jason Edwards draws on Michel Foucault's work to understand the relationship between war, power and politics. He argues that Foucault leads us to a very particular reading of politics as war by other means, one which avoids the reductive essentialism of other leading accounts (especially those grounded on Carl Schmitt's celebrated friend-enemy distinction). In Edwards's reading, Foucault recasts and revitalizes the idea of politics as an alternative to war. Finally, Bob Brecher raises an especially urgent question in the context of the on-going 'War on Terror': is there such a thing as political terrorism? He argues that there is not because no actual terrorist group pursues violence as 'an end in itself.' Each of the three authors challenges the contemporary understandings of what war means.

The second section of the book is focused on changing means of pursuing security in the context of the new wars. Avery Plaw examines the application of existing international law to a new, contentious and increasingly prominent instrument of national security—the targeted killing of terrorists who are thought to pose an urgent threat to civilians. He focuses his analysis on one relatively well-documented case, a US Predator targeting operation in Yemen in 2002. He concludes that while there appears to be a plausible case that this operation may have been legal under existing law, examination of it raises issues that cast doubt on the legality of some other US operations. In particular, he suggests the need for open US recognition of its own policy and suggests a variety of reforms to ensure that operations conform with, and hence uphold, legal standards for the conduct of war. Benjamin Rampp next presents a critique of the contemporary concept of human security, illuminating how it has lost meaning and hence protective power through excessive expansion. He argues forcefully for a process of desecuritisation, focusing in particular on narrowing (or un-broadening) the concept. Stuart Murray then exposes the obsolescence of conventional conceptions of diplomacy as the 'engine room of international relations.'[29] This image, he argues, is not only inaccurate but undermines diplomacy's capacity to contribute to the maintenance of national and human security. Fortunately, this conventional account has been challenged from several directions by new and emergent schools of diplomatic thought. Murray identifies, distils and synthesizes these schools of diplomatic thought thus revealing how to renew the study and practice of diplomacy as a means of preventing and limiting war in its contemporary guises. All three authors examine and recommend new strategies for improving the prospects of human security in the face of new and emerging threats.

The third section of the book examines changing representations and experiences of war. Mustafa Serdar Palabıyık examines how concepts of war continuously evolved throughout Ottoman history, re-shaped by leading public officials and intellectuals. He shows how the transformation of the representations of war evolved with the historical conditions and circumstances of the Empire. Pamela Chraibieh Badine's chapter turns from the changing representation of war to the enduring memories of it, how these memories can themselves present a continuing obstacle to peace, and how they can be finally transformed in a process of reconciliation. Specifically, she examines the case of Lebanon, a country that has especially suffered in recent decades with wars both civil and international in character. Her chapter presents a micro-analysis of the work—the successes and the failures—of five specific groups working with victims of Lebanon's various wars, devoted to helping both individuals and the larger society to live with their memories, and to find the strength in them to fashion a new politics. Finally, Tim Markham explores the different strategies that reporters are bringing to the representation of contemporary wars. He draws attention in particular to the different ways correspondents' experiences are shaped by the physical constraints of pool and/or embedded reporting, and how their work is evaluated through two increasingly influential symbolic economies (which he terms esotericisation and ambivalence). Markham calls for deeper self-reflection among reporters in order to understand and control these new influences, and kicks off the process himself by identifying four ways that the contemporary technology of correspondence from combat zones influences the product. All three authors devote their attention to evolving patterns in the representation and remembrance of war.

The fourth and final section of this book examines sources, strategies and challenges of the new forms of warfare. In his contribution Timothy D. Hoyt examines the political and sociological logics which foster terrorist movements. Specifically, he considers the case of the Irish Republican Army (IRA) which seems to re-emerge each generation 'like a phoenix from the ashes.' Hoyt concentrates in particular on the sources of its resilience, suggesting means by which the power and violence of such groups could in future be diminished and controlled. Next, in a strategic approach to the recently ended 30 year civil war in Sri Lanka, Graeme Goldsworthy, Toby Chesson and Erica Pasini illuminate internal mechanisms of transformation in the conduct of low intensity warfare. They focus chiefly on the use of landmines in Sri Lanka, and raise the question of whether the recent international landmine convention is likely to improve or exacerbate the problem. They offer reason to think that improvement is unlikely in such cases, and thereby challenge the reader to reconsider the treaty's mechanisms for influencing low intensity conflict like the Sri Lankan civil war. Finally, a fundamental question raised by new, unconventional forms of warfare, according to Bård Maeland, is the challenge of boredom. Of course boredom, as Maeland notes, has always presented a challenge to soldiers; indeed, a traditional description of a soldier's life is '5% horror and 95% boredom/waiting.'[30] But because contemporary conflicts tend to be 'low intensity' by comparison with traditional conflicts, the period between engagements has stretched, intensifying the challenge of resisting ennui and remaining alert and prepared. Maeland examines both empirical and anecdotal evidence to better understand the challenge posed to modern soldiers and to suggest strategies that military leadership can employ to resist this subtle scourge of modern warfare.

4. Conclusion

Together, these chapters make a substantial and important contribution to understanding the ways that war has transformed in our own time as well as in some prior eras. They shed considerable light on the special challenges posed by the new wars, and make some suggestions as to how it may be possible to extricate ourselves from them. They also, however, reveal the magnitude and complexity of the subject area, and in particular the range of disciplines it necessarily calls into play. This introduction began by drawing on some heroic individuals' efforts to come to grips with the metamorphoses of war—including those of Münkler, Van Creveld, Tilly and von Clausewitz. But it would be improvident to rely solely on the work of great individual scholars. There is an urgent need for collaborative work bringing multiple disciplines to bear on the on-going transformations of war—all the more so in view of the increasing toll it has taken on civilian lives and essential public institutions (including the state) in recent decades. It is to instigating such work that this book is ultimately devoted.

Notes

[1] H. Münkler, *The New Wars*, trans. P. Camiller Polity Press, Cambridge, 2005, pp. 1, 12-13.
[2] Ibid., p. 4.
[3] Münkler himself introduces the language of 'state-building' and 'state-disintegrating' wars: Ibid., p. 8.
[4] D. Kennedy, *Of War and Law*, Princeton University Press, Princeton, 2006; C. Greenwood, 'Historical Development and Legal Basis' and 'Scope of Application of Humanitarian Law,' *The Handbook of Humanitarian Law and Armed Conflict*, D. Fleck (ed), Oxford University Press, Oxford, 2008, pp. 10-45.
[5] C. von Clausewitz, *On War*, trans. M. Howard and P. Paret, Princeton University Press, Princeton, 1989, p. 87.
[6] Ibid., p. 75.
[7] Ibid., p. 81.
[8] Ibid., p. 75.
[9] Ibid., pp. 581 and 584.
[10] Ibid., pp. 583-584.
[11] M. van Creveld, *The Transformation of War*, Free Press, New York, 1991, p. 307.
[12] Ibid., pp. 25, 20-22.
[13] Ibid., pp. 20-21.
[14] Ibid., p. 20.
[15] Ibid., p. 73.
[16] Münkler, op cit., p. 3.
[17] Ibid., p. 1.
[18] C. Tilly, *Coercion, Capital and European States*, Blackwell, Oxford, 1990, pp. 197, 201-202.
[19] Ibid., pp. 202-203.
[20] S. David, *Catastrophic Consequences: Civil Wars and American Interests*, The Johns Hopkins University Press, Baltimore, 2008, p. 7.
[21] M. Kaldor, *New and Old Wars: Organized Violence in a Global Era*, Polity Press, Cambridge, 1999, p. 216; C. Clapham, *War and State Formation in Ethiopia and Eritrea*, The Global Site, University of Sussex, England, 2000, Accessed May 1, 2008, <http://www.theglobalsite.ac.uk/press/010clapham.pdf>; C. McInnes, *Spectator Sport-War*, London, Lynne Rienner, 2002.
[22] Kaldor, op. cit., pp. 5-11; see also, for example, S. Huntington, *The Clash of Civilizations and the Remaking or World Order*, Simon & Schuster, New York, 1996, pp. 246-98.
[23] Kaldor, op. cit., p. 8.

[24] Creveld, op. cit., p. 212.
[25] Creveld, op. cit., pp. 25-32.
[26] Crevld, op. cit., pp. 23-25.
[27] Kaldor, op. cit., p. 8.
[28] B. Hoffman, *Inside Terrorism*, Columbia University Press, New York, 1998, p. 134; also see pp. 131-135.
[29] S. Murray, 'Towards an Enhanced Understanding of Diplomacy as the Business of Peace,' *The Metamorphosis of War*, A. Plaw (ed), Rodopi, Amsterdam and New York, NY, 2012, pp. 99 and 100.
[30] B. Maeland, 'From Manifest Degradation to Latent Anticipation: Military Boredom in the First World War and Afghanistan,' *The Metamorphosis of War*, A. Plaw (ed), Rodopi, Amsterdam and New York, NY, 2012, p. 243.

Bibliography

Clapham, C., War and State formation In Ethiopia and Eritrea. The Global Site. University of Sussex, England, 2000. Accessed May 1, 2008. <http://www.theglobalsite.ac.uk/press/010clapham.pdf >.

Creveld, M., *The Transformation of War.* Free Press, New York, 1991.

Greenwood, C., 'Historical Development and Legal Basis' and 'Scope of Application of Humanitarian Law.' *The Handbook of International Law in Armed Conflict*. Fleck, D. (ed), Oxford University Press, Oxford, 2008.

Hoffman, B., *Inside Terrorism*. Columbia University Press, New York, 1998.

Huntington, S., *The Clash of Civilizations and the Remaking of World Order.* Simon & Schuster, New York, 1996.

Kaldor, M., *New and Old Wars: Organized Violence in a Global Era.* Polity Press, Cambridge, 1999.

Kennedy, D., *Of War and Law.* Princeton University Press, Princeton, 2006.

Maeland, B., 'From Manifest Degradation to Latent Anticipation: Military Boredom in the First World War and Afghanistan.' *The Metamorphosis of War.* Plaw, A. (ed), Rodopi, Amsterdam and New York, NY, 2012.

McInnes, C., *Spectator-Sport War*. Lynne Rienner, London, 2002.

Münkler, H., *The New Wars*. Trans. Camiller, P., Polity Press, Cambridge, 2005.

Murray, S., 'Towards an Enhanced Understanding of Diplomacy as the Business of Peace.' *The Metamorphosis of War*. Plaw, A. (ed), Rodopi, Amsterdam and New York, NY, 2012.

Tilly, C., *Coercion, Capital and European States*. Blackwell, Oxford, 1990.

von Clausewitz, C., *On War*. Trans. Howard, M. and Paret, P., Princeton, Princeton University Press, 1989.

Avery Plaw is an Associate Professor of Political Science at the University of Massachusetts at Dartmouth. He is the author of *Targeting Terrorists: A License to Kill?* and *Frontiers of Diversity: Explorations in Contemporary Pluralism*.

Axel Augé is a sociologist and Senior Research Fellow at The Centre for Saint-Cyr Military Academy Research (CREC). He's currently teaching Military Sociology at Coëtquidan Saint-Cyr Military Academy (France).

Part I:

New Concepts of War and Terror

Fighting For Peace: From the Social War to Armed Democracy

Nick Mansfield

Abstract
The canonical theories of war divide between the account linked with Hobbes and Kant, which supposedly sees war as the opposite to civil society, and that of Clausewitz which sees war as continuous with politics. This chapter analyses these thinkers to show that their accounts were (often unintentionally) more ambiguous than is usually thought. The consequence is that modern and postmodern accounts of war from Freud to Derrida, which see war as paradoxical, both incited and refused by society, offer compelling insights into what war has been for western modernity. This account of war as 'double' has major consequences for our thinking about the relationship between war, on the one hand, and democracy and human rights, on the other. Democracy, for example, can be seen both to require and refuse violence and war.

Key Words: War, Derrida, deconstruction, Levinas, democracy, human rights, Carl Schmitt.

*****

From the great revolutions of the late 18th century to the Universal Declaration of Human Rights, the great liberal breakthroughs of the modern era all arose through, as a consequence of or as a reaction to war. No era can match the modern era for violence: from the outrage of slavery to the absolute outrage of the genocides of enslaved, indigenous and minority populations; from the invention to the plundering of the 'colony,' from the anguish of the police state to the insidious manipulations of biopower, no era has been as haunted as the modern by international, internecine or social war. The simultaneous progress and atrocity of this era cannot in every case simply be set as opposites to one another. The liberal and violent commonly emerge together in inextricable relationship with one another as part of a single double operation. Authoritarian military systems collapse into democracy. Wars are fought on behalf of human rights. Yet we insist on the straightforward opposition between these things: between war and violence, on the one hand, and all the things routinely opposed to them, on the other—peace, love, friendship, society, humanity, progress, law and order and so on and so on. The relationship between these things tends to be framed moralistically or rhetorically. We revile war and love peace. We fear violence and trust in friendship. The polarisations in which we invest so much in our campaigns against the ever-threatening renewal of war are themselves unthinking, even bullying. Moral preference over-floods theoretical precision. Of course, war is a moral issue, and our priority should always be to imagine its possible reduction to zero. Yet, we cannot afford to do this too quickly. If we imagine that the relationship between war and civility in general is uncomplicated, or that our insistence on simple categories and oppositions must be ruthlessly adhered to if we are to achieve our goal, or indeed that it can deliver our goal, rather than re-plunge us into a cycle of horror and despair, we risk remaining entangled in the complex and obscure situation in which war may indeed have been the thing that has delivered us the democracies and the human rights we deserve, even as it remains the greatest challenge to them. War, in short, both protects, even advances our freedoms while threatening and mutilating them. The freedoms and rights we want to preserve from the violence and threat of war may indeed be risked without war to protect them. How can we have everything? How can we reduce war while preserving the things for which wars are fought? The aim of this chapter is to propose a way of thinking about war and what is other to it which does justice to this complexity. Only by understanding our relationship to war in all its complexity can we imagine moving beyond narratives that are only rhetorically satisfactory.

The classical accounts of war divide into two streams. On the one hand, there is the position taken by Hobbes and Kant that, in different ways, sees war as the degrading natural state of humanity from which we are trying to escape when we institute organised society and sovereign authority. On the other hand, there is the West's most influential and cited understanding of war in Clausewitz, that war is the continuation of politics by other means. These accounts survive today, of course. Behind the intensifying conservative social policies dominant in western countries since the 1980s, we detect the Hobbesian idea that the social is perched precariously on top of a chaos that threatens constantly to suck it down into interminable and irredeemable social violence. Here, war looms at us from beyond the frontier between the social and pre- or anti-social. The forces of law and order must themselves take up arms to keep out violent disorganisation. So pressing is this social war that we must fight it. The re-casting of social disputes as wars is not a mere rhetorical gesture. Rather, it partakes of a fully developed if only partly revealed theory of society wherein war defines the limit of the social.

The Clausewitzian position is, incredibly enough, now ascendant on the other 'critical' side of the political divide. The idea that war and policy are now seamlessly co-ordinated is seen by a host of radical critics from Foucault to Badiou and Antonio Negri as a way of revealing the violence of the liberal global agenda. The classical accounts endure then as the defining ways in which the relationship between war and civil society is conceived.

These accounts are more complicated and fraught than they are usually thought to be, however, and, although they seem to rely on simple positions—war either as the opposite to the social, or continuous with it—they prove in fact to be far more unstable and ambiguous. In Hobbes, human beings are seen to be motivated by desire, but, more importantly, they act in order to maximise their freedom to desire in the long term. What motivates us then is not simple desire, but our will-to-desire, our desire to desire, our demand not that certain measurable and knowable desires can be fulfilled (thus bringing the process of desire to an end), but the freedom by which our desire, whatever it may be, can always be imagined as horizon-less and endlessly achievable. Initially, human beings believe that this freedom of desire will be available in the state of nature, where there is no restriction on our grasping behaviour. However, it becomes clear that in nature, there can be no regulation of privacy, thus making the right to property impossible, nor can there be any measurement of equity, thus ruining any chance of there being justice. Desire cannot be free where there is no protection of your right to hold onto what you want. Nor can your desire be respected as yours and as equal to anyone else's without some basic level of arbitration that would guarantee that your desire will not simply be overwhelmed by someone stronger or with some other strategic advantage. War is Hobbes's name for this state of nature, where competition is unfair, unregulated and blind.

Perceiving that the freedom of desire is not possible within nature, human beings band together to institute civil society and sovereign authority. By surrendering some of our freedom to a greater authority, we produce the social order by which our desire becomes both attainable and securable. Overseen by an incontestable authority, human beings can suddenly exercise the freedom they naturally seek. By excluding war from human interactions, we can suddenly attain the very thing that we sought in war: our freedom of desire. However, this idea is complicated by the way in which Hobbes presents this process. This was a complication of which he was aware, but that he thought he could circumvent. The problem is that what Hobbes presents as a narrative is not meant to be taken literally as such. Hobbes's account reads as if the move human beings make from the state of nature to living in a society governed by sovereign authority is a simple two-stage process, and that these stages are to be read literally as discrete historical phases. This is, however, not the case. It is not simply possible to identify this achievement as a simple story of evolutionary progress. The state of nature continues to coincide with organised civil society, in three ways: firstly by the fact that whole peoples still live in a wild state (Hobbes's example is the indigenous peoples of the Americas); and secondly, the relationship between states remains comparable to that between individuals in the state of nature, and no overarching mechanism exists to oversee this inter-state anarchy; finally, and most importantly for our discussion, the state of nature is not simply something human beings leave behind as they enter into ordered social relations: it remains permanently in the background, ever pressing against our civility, ever-threatening, and always requiring to be policed and controlled by a larger authority. In other words, the state of nature endures permanently as the underside of ordered social relations. A constant tension persists between our will to obey sovereign authority and our natural inclination towards competition and self-interest. This perception endures everywhere in the social policies that descend from Hobbes. The state of nature is not some remote and archaic curiosity. It exerts serious pressure on us constantly, and must be consistently and ruthlessly policed. We cannot be left to ourselves, or else brutal competition will burst out, easily overwhelm the institutions of civic order and plunge us into a chaos of radical individualism and vicious competition.

We have seen already that the point of instituting sovereign authority was to make possible the realisation of desire—what was sought but unachievable in the state of nature. Civil society is not the place where nature is suppressed, but the very place where our natural desire is achievable. Nature presses on it constantly, accompanying it permanently. The fact that society is the place where desire is to be achieved means that the very thing that defined nature also governs social relations. Society then becomes the place where our nature can be fulfilled: in other words, it achieves war's goals. In Hobbes, therefore, war is sewn into the lining of social relations as their meaning, the violent thing of which they are the legitimate version. Society is war made acceptable. In short, Hobbes's account of the relationship between war and civil society cannot simply be seen as one of polarisation. War and the social always occur together. Society achieves war's goals, and war constantly haunts the social as its permanent accompaniment. War is not society's alienated opposite but its constant double. Society fulfils the ambitions of the war it must exclude and abominate. War is simultaneously facilitated and repressed. The successful operation of this system requires, however, that the confusion of this argument needs to be denied. Hence the continual insistence on the idea that chaos and desire, on the one hand, and law and order, on the other hand, are polar opposites. A consumer economy and a culture of power where desire and individual agency are constantly fetishised is continually accompanied by this suspicion and abomination of the excesses of desire: desire is encouraged, even compulsory, but remains dangerous, and the thing where individuals most easily expose themselves to sanction, from ridicule to imprisonment, if they exhibit their desire too excessively or explicitly, or if their particular desires just happen to be the focus of abuse or the current moral panic. In sum, Hobbes's account imagines the polarisation of two things that it represents as antagonists that are, in fact, mutually defining, entangled, inseparable and inter-penetrating.

The same complexity arises in a different form in Kant's account in his argument for an international covenant which will make perpetual peace possible. For Kant, the natural state of humanity is also one of war. This state is one of lawlessness. A life without law is implicitly violent to Kant. Indeed, people living without law are implicitly at war, even if they perpetrate no violence against one another. As with Hobbes, if unregulated violence is simply just possible, then human beings are at war, even if violence does not occur. This is the first indication of the complexity of Kant's description of war and peace, one that we have inherited and that in the course of the last two centuries has been put into practice frequently. War is the state of lawlessness, whether there is or is not violence. Peace is the state of lawfulness. Peace here becomes dissociated from non-violence and war can occur without violence, and can simply be the mere possibility of violence, or even just the lawless act. This dissociation means that there is a definition of peace more important than non-violence, a peace, therefore, that might not be incompatible with violence. Concomitantly, there is nothing to stand in the way of a peace that may depend on violence, and only be possible by way of it.

It is reason that constructs the morality to which human beings must adhere if they want to raise themselves out of the state of nature. Human duty is to make the edicts of reason practical and practically achievable. Politics should conform to morality. 'All politics must bend its knee before the right,'[1] Kant argues. It is the duty of human beings to implement the moral priorities of reason. Reason will lead to 'perpetual peace,' according to Kant. This is not because reason is implicitly conducive to what we normally understand as peace. As we have seen, in Kant, peace may not be the thing we normally understand it to be. Indeed, the reason that duty leads inevitably to peace is that obeisance to duty *is* peace. The drive to institute the moral law is 'the universal will, given a priori ...'[2] To live in a state governed by law is to be implicitly in a state of peace. The state of war is the state of lawlessness, whether or not it leads to violence. Again, we see in Kant the dissociation between peace and non-violence. Again, the consequence of this could be that there is nothing more peaceful than violence, if violence is deployed to protect or advance the law. No-one needs to be reminded that the pursuit of peace by way of violence, or through war on a scale previously unimaginable, has been a common project in post-Enlightenment history.

Consistent with this implication here, Kant goes on to argue that peace may only be achievable after and through war. Kant writes that nature's 'aim is to produce a harmony among men, against their will and indeed through their discord.'[3] To Kant, peace is not a state that can be realised on its own. Human beings must learn through war that peace is their destiny. This develops in several ways. Firstly, war drives defeated populations into the earth's most remote corners where they learn that only co-operation will allow them to survive. Secondly, and most importantly, human groups must organise themselves into co-ordinated groups simply in order to fight wars. War would not be possible if at the tribal, communal or national levels, people resisted the impulse towards organisation and collaboration, in other words, an incipient peacefulness with one another. War then is its own worst enemy: it produces its very opposite in the course of its own operation. War is the factory of peace. The inverse conclusion does not seem so bright: if war produces peace, it means that peace requires war as the passage through which we must progress in order to attain our natural, lawful state of orderliness and co-operation.

In both Hobbes and Kant, therefore, we do not see the simple alienation of war and peace from one another. In Hobbes, social order is instituted in order to control or exclude war, yet this order's fundamental aim is to achieve what it is in human beings—desire—that gave rise to war in the first place. War remains social order's hidden meaning. In Kant, war becomes identified with lawlessness, and peace with duty. What we might conventionally understand as war—organised and official, lethal violence—could then become what it has been so often in post-Enlightenment European history, the ostensible means to peace. This idea is reinforced by Kant's argument that war itself incites in human beings the desire for co-operation, the very co-operation that will teach us the advantages of peace. In both thinkers, peace and war are entangled with one another even as they are nominally treated as opposites. Despite the proclamation of a peace as the end of social processes, something that arises *beyond* war, peace never seems able to purify itself of the war to which it owes its value, and its evolution.

The complex entanglement of war and peace emerges in another form in the most famous view of war in the Western tradition, Carl von Clausewitz's statement that 'war is a mere continuation of policy by other means.'[4] Clausewitz's argument counters two alternative ways of thinking about warfare. The first of these is that war is an aristocratic art, in which an élite warrior caste plays out its traditional rites of valour and honour. The second is that war represents the point where diplomacy fails, so that relations spin out of control and fall into an abyss, exposing us to chaos and violence. To Clausewitz, war is merely another way in which purposeful human groups rationally pursue their goals. War is primarily a tool of State policy.

Clausewitz's argument is the most cited account of war in the post-Enlightenment period. As we see in Jason Edwards's chapter in this volume, in *Society Must be Defended,* Foucault has famously reversed Clausewitz in a spectacular rhetorical move in which he proposes—in an argument that clarifies the evolution of biopolitics—that social power is the perpetuation of unfinished and forgotten wars. Many other contemporary analysts of war refer to Clausewitz's argument as proof that recent wars are merely alternative methods used by cynical governments to pursue the global neo-liberal agenda. This is not a complete or satisfactory representation of Clausewitz's account of war, however. Clausewitz was writing at the time when mass revolutionary movements were first having an impact on continental politics. Valmy had shown that popular force could trump aristocratic tradition on the battlefield. The physical might of the collective energy of the people was a startling new factor in modern warfare, and needed to be understood. This mass energy was brutal and blind to Clausewitz. It did not share the rational clarity of the directed policy that gave war its cool and meaningful purpose. The actions of the people were a spasm of mighty but mere partiality. They acted not to achieve specific and knowable goals, but out of general inherited antagonisms. They did not seek merely to subject the enemy to their will, but to annihilate them. To Clausewitz, this energy was an irresistible force, one that needed to be taken into account in the planning of wars, but it needed to be controlled and directed. It could be used as the vehicle of policy, if only the right means could be found to bend it to State purpose. The lynchpin that could join élite purpose and mass energy was the individual genius. Napoleon, of course, looms as the archetype.

In Clausewitz's account, therefore, mass force is harnessed for official goals by the charismatic leader. What is important here is the co-ordination of two drives: one in which the focussed here-and-now logic of the practical operation of the State combines with the apocalyptic drive of the people. As we will see below, this foreshadows the distinction made by Walter Benjamin between mythic violence (the violence of State purpose) and messianic violence (the violence of unknowable revolutionary change), a distinction problematised and developed by Derrida. Suffice it to say that, at the dawn of the liberal democratic era, war is perceived by the West's most famous war theorist as not simply the execution of fixed policy goals, but the uneasy and potentially dangerous forced co-ordination between these goals and something else, the energy of the people. Clausewitz's account then is not one of the directed drives of a well-oiled and sweetly operating social machine. It is one in which the totality of society is put into action—state purpose and popular force—in some kind of collaboration but one of tension, contradiction and risk, where the clash between top-down purpose and bottom-up energy remains live and potentially explosive. It is society in action, but as a problem. This account is clearly disdainful if not fearful of the 'people.' The image of a blind mob guided by sober leadership would now function in politics as a kind of conservative fantasy, one that is not easy to avow, especially in states where élite politicians need to placate or fool a populist media. However simplistic or chimerical it is, Clausewitz provides an image of the complex entanglement of purpose and force in democratic politics, one that our wars of liberal Enlightenment still try to enact. Purpose presents itself as liberal and rational, yet it remains in thrall to, and at times even adores, the violence of uncompromising aggression, seen as a vehicle of liberty, of the teaching of lessons and of progress. This is one of the ways in which democracy is still practiced, as the uneasy co-ordination of arcane theoretical intention and mass appeal. The Iraq war showed us this particular version of democracy: an esoteric cabal finding the perfect partner in a compliant mass media and popular culture.

In these various accounts, what we have seen is a series of attempts to present war in a simple relationship with what is other to it, civil society. Yet, these accounts prove much more complicated and ambiguous than they first appear. War either accompanies society as its permanent double, becomes the meaning of or even the means to peace, or else it puts society to action as the uneasy co-ordination of contrary fractions. In each of these instances, war conditions the social, defining its purpose, driving its advancement or enacting its meaning. In short, the achievement of sociality that we would see as most under threat from war, requires war.

Yet there have been attempts to deal with this complex and ambiguous situation, and indeed at least one of these attempts—Freud's—has been widely taken up. In this view, war and peace are neither separate from, nor continuous with one another, but in a relationship of ambivalence, unresolved contradiction or doubleness. Freud argued that war involves the re-emergence of primitive violent impulses we thought had been superseded by civilisation, but that in fact had simply been repressed. Indeed, according to this account, the processes of repression on which sober civil relations depend do not reduce our violent impulses, as much as intensify them by way of an ever tightening constriction of them that must find expression somewhere. In other words, the attempt to control unconscious violence compounds and strengthens it. The control of and incitement to violence then become part of a single process. The more civilised we are, the more intense our inner violence will be. It should have been no surprise then, when the cosmopolitan civilisations Freud thought of as the highest achievement of humanity, exploded in the most atavistic nationalist violence in 1914. Theorising this event, which brought on a moral crisis for the young science of psychoanalysis, Freud was able to argue that what war showed was the ambivalence at the heart of all human relationships: civilisation collapsing into violence, love into murderous resentment, identification into loathing. This idea of war as the outpouring of sublimated subjective violence entered modern consciousness, and has become part of our psychological common sense, like so much of the Freudian legacy. We can see it in the a-historical way in which the peace movement, for example, sees war not as a phase in the unfolding of specific political events, but as the expression of elementary forms of social, cultural or gender-based violence.

In the Hobbesian and Kantian accounts, war was putatively the opposite of civil society; in Clausewitz's, the continuation of it. In Freud's account, the relationship is acknowledged to be far more complicated. The social incites war while repudiating, even anathematising, it. The relationship is contradictory, ambivalent and irreducibly problematic. War emerges by way of the very social relations that would seem opposite to it, and that we like to think we are determined to protect from war. The relationship between war and civil society is not one of opposition or identity, but of dangerous doubleness and obscure entanglement.

This way of thinking about war has been picked up by a variety of modern and postmodern thinkers, from Bataille to Deleuze and Guattari. It is in Jacques Derrida's discussion of war that this thinking of war and the social in terms of doubleness reaches its most mature expression, however, and for this reason, I want to dwell on it a little more fully. Derrida's account of war proceeds by way of the deconstruction of the two key binary oppositions on which a clear definition of war would seem to depend: on the one hand, the distinction between war and peace; on the other, between friend and enemy. The first deconstruction emerges through a reading of Emmanuel Levinas' discussion of war in *Totality and Infinity*, and the second through Derrida's reading of Carl Schmitt's *The Concept of the Political.*

Derrida's discussion of Levinas is a commentary on the latter's tracing of both war and peace back to their common source in the relationship to otherness anterior to all things. Levinas does not see war and peace as opposites to one another. Both war and peace arise as possibilities of human intersubjectivity on the same terms, and in the same direction. 'Only beings capable of war can rise to peace,' Levinas writes.[5] This is because, to Levinas, human being and everything it encompasses—subjectivity, agency, historicity—all emerge only after a prior opening towards the otherness that calls us into being, the fact that we always arrive later than something other that must always precede us. We emerge only in relation to a previously existing otherness to which we are responding, before we have any choice in the matter. Our openness to this otherness is summarised as our relationship to what Levinas calls 'the face,' the absolute and ineluctable degree zero of our orientation to the difference of otherness, which exists prior to that relationship having any content. Before I know who or even what you are, your face opens to me in the simple incipient possibility of relationship. In this opening arises both the possibility of war and of peace. Levinas writes: 'War … presupposes[s] the face and the transcendence of the being appearing in the face.'[6]

To Derrida, Levinas' argument is the inverse of the one we have seen in thinkers like Hobbes and Kant. Where they would see war as pre-existing peace, and peace as always threatened by the re-emergence of war, Levinas argues that openness to the other, the very logic of peace, pre-exists war. This would seem to say that peace is larger than, and encloses war, and thus always implicitly triumphs over it, but to Derrida, there is a certain horror to this logic. Levinas may have effectively deconstructed the opposition between war and peace, but Derrida wants to insist on the dangerous, unstable and infinitely problematic situation this produces. If Levinas were right, war would always be notionally or residually peaceful. It would not, therefore, be something to abominate and refuse, but something to which we could be resigned as simply another denomination of human relationship. Thus, to Derrida, it is not possible to simply exclude war from the social by definition, nor is it possible to accept it as an inevitable part of our relationships. It both arises from within our sociality and is the scourge of it. It opens as part of the opening of all that is good about us. Yet it threatens to ruin it as well and we do not accept it. Sociality is thus irreducibly double and problematic. We can neither simply exclude war nor resign ourselves to it.

Derrida also deconstructs the other binary opposition on which the identity and meaning of war would seem to depend: the opposition between friend and enemy. Carl Schmitt had argued influentially that the definition of a political group depended on the identification of a shared enemy, and the real possibility of war with them. To Schmitt, this enmity was not the result of any aspect of the enemy, their ethnicity, for example. It was not personal but structural, not a response to the other's foreign nature, but the construction of a role for them in the self-identification of the group. The relationship with the enemy-other, therefore, is not a result of personal antagonism. Indeed, such public enmity could co-exist with private friendship. The problem with this contradiction, as Derrida points out, is that the co-existence of your private friendship with public enmity is only possible in a world where the distinction between public and private already exists, but this is the very distinction that the identification of the enemy brings into existence for the first time. In other words, the identification of the enemy refers to a pre-existing distinction between public and private that only it can bring into existence. In short, Schmitt's distinction between friend and enemy cannot ground itself.

This dense argument—the deconstruction of the oppositions between war and peace, on the one hand, and friend and enemy on the other—is not a mere intellectual game. What I want to argue is that our normal ways of understanding these relationships are inadequate to at least our current historical circumstances. The idea that war and peace are mutually exclusive or identical with one another fails to grasp the complexity of the situations of war with which we have to deal. Indeed, I want to argue that it is to a deconstructive Derridean logic that we must look if we want to understand the doubleness of the wars in which we find ourselves. Our conventional accounts attempt to advance a purely moral logic: firstly, the programmatic idea that peace is separable from war and redeemable from it; the other, the sophisticated one that the corruption and cynicism of our apparently peaceful social institutions condemns us to war. Yet the wars we are involved in are ones in which the distinction between war and peace and between friend and enemy are not at all clear. We fight wars on behalf of civic values, like democracy and human rights, for example, that we consider to be the sweetest fruits of peace, which use atrocity, rape, arbitrary detention and collective retribution as methods, while the disciplines of warfare also end in the muting if not savaging of human rights at home. Similarly, our domestic politics have long been a zone in which the ordinary administration of social policy is imagined as warfare: we have had wars on poverty, on drugs and on crime, history wars and culture wars, and more recently, with the war on terror, this logic has been projected across the whole of the earth.

Similarly, the distinction between friend and enemy is also irreducibly problematic: a figure like Saddam Hussein, for example, was both a bulwark against Iranian fundamentalism whose political crimes had to be explained away, and the re-incarnation of Hitler, to be overthrown and executed for the same crimes. A figure like this, however, is simply a sensational example of a much deeper problem: in asymmetrical warfare in particular, it's not clear who in the civilian population you are fighting for and who you are fighting against. This is not simply the result of the inability to discern one from the other. There is a loss of separation between these categories to the point where civilians who are clearly non-combatants become targets of a generalised military terror using torture and rape, whose aim is to keep whole populations in a state of intimidation and neutralisation, designed to liberate them. We must attack you in order to defend you. The Vietnam era statement that the village needed to be destroyed in order to be saved is not a cataclysmic absurdity, but evidence of the deepest problematic of war.

The loss of difference between war and peace has been much commented upon. What we must not forget, however, following Derrida's commentary on Levinas, is that this relationship remains contradictory. Wars may be fought to advance democracy, for example, but only by radically unsettling, even transforming, democracy in the process. What we have here is not a false democracy, as so many critics would say, but something far more challenging: a democracy built on the continuous management of public freedom by violence, not a violence that functions as a transition phase to be worked through, nor one that merely operates at the perimeter of the social, but as a permanent accompaniment to the democracy it both facilitates and threatens.

The deconstruction of the relationship between war and peace, or friend and enemy does not reduce these differences to identity. The distinction becomes unfixed and contingent, but difference remains, as an irreducible disjunction. Wars are fought to expand democracy and human rights that mutilate democracy and human rights in both the theatre of war and at home, and that are resisted in the name of democracy and human rights. As Bob Brecher points out in an accompanying chapter, a war fought against terror inevitably becomes terroristic itself.[7] This is not irony on the part of history, but must be seen as a fundamental clue to what democracy and human rights are for us, and what war is. Wars are provoked by democracy as part of its operation. Democracy is an armed concept. It does not simply use warfare, nor is war simply the transition we must go through in order to reach democracy. Democracy requires war. It is to this problem that we now turn.

Derrida once famously wrote that there was no law which did not at some point rely on the idea that it would need one day to be enforced, by violence if necessary. This argument can inform our understanding of the relationship between war, on the one hand, and democracy and human rights, on the other, and is thus worth investigating in some detail. To Derrida, law represents the attempt to realise the call of justice. Justice here is not to be understood as a measurable social or legal equity, nor as the most appropriate calibration of social order in a particular historical context. Undefinable, immeasurable and unreachable, justice is the asymptote of responsibility to the other. In this way, justice remains perpetually in excess of any human action, because, following Levinas again, this openness is constitutive of subjectivity and agency. Since agency depends on this openness to the other, justice cannot be finally fulfilled. No action can be completely without reference to it, nor, in turn can any action so completely satisfy the call of justice that the latter will be finished. Law, then, only makes sense in relation to the ever-receding horizon of a justice that has called it to action, but that, in turn it can never finally put to rest. The call to justice will never be surpassed. Justice licenses law, but continues to exceed it, ever calling for law to be renewed, extended and *re-formed.* This explains the mechanism by which legal change remains an interminable historical process.

The relationship between law and justice is not, however, an unambiguous one. Derrida maps his discussion of the relationship between law and justice onto Walter Benjamin's discussion in 'Critique of Violence' of the contrast between two types of violence: mythic violence, that maintains social and political order, and messianic violence, which smashes, overturns and re-founds political orders. The latter is an account of the revolutionary impetus to challenge and re-make the social order. Messianic violence which makes sense in relation to the potentially cataclysmic founding of any social order will always be in excess of mythic violence. Mythic violence will always be in some way oriented towards the messianic violence with which it will inevitably have a contradictory relationship. The situation that allows mythic violence to operate can only be brought into being by an event of messianic violence. Yet, at the same time, messianic violence will always threaten the mythic with the possibility of a revolutionary spasm that will completely overturn it. The messianic institutes and licenses the mythic, while remaining the greatest threat to it.

The relationship between law and justice is the same. Justice institutes law, and law only takes on meaning in relation to it. On the other hand, justice will always threaten law, not only with the endlessly open historical process of adaptation, reform and renewal, but also with the possibility of complete disruption, abandonment and abolition. Justice then both makes law and threatens it. It both guarantees the system of legal political organization and the institution of right, and remains the greatest danger to it. Justice, like messianic violence, is simultaneously the image of human aspiration to change and social renewal, and the promise of apocalypse. It thus bears with it the promise of systematic social order (and the disciplinary violence this may require), and revolutionary social change (and the open-ended content-less violence this may incite).

It is too simple, therefore, to see law and justice as separate or even different to one another. Law has no meaning without reference to justice, and justice in turn, because of its hyperbolic nature, can never be achieved in the world in itself. Justice only appears in and through the acts of law supposed to aspire to it. It is, in fact, not a separable entity, but merely the tendency to self-overcoming in the law itself. The violence of the enforcement of law, then, takes potentially two forms: one of social repression in the name of a known and constituted order, and the other, the violence of the undefinable, and putative imaginary absolute, to which that constituted order nominally refers, and which it had hoped to honour and implement in the first place. In particular historical circumstances, since law and justice are not separable but part of a single double nature, the violence of promulgated law and the violence of indefinable and unconditional justice easily become confused, till tropes like 'the right,' 'sovereignty,' even 'justice' itself become simultaneously knowable political quantities and intimidating, immeasurable and ultimately potentially lethal, if easily excused, rhetorical gestures.

The situation with democracy and human rights is similar. Democracy is, on the one hand, a type of fixed political organization, and, on the other, an image of radical aspiration, which that political organization is supposed to exemplify. This latter, which Derrida in his late work, repeatedly evoked as 'democracy-to-come,' remains the image to which all nominally democratic systems aspire, of an undefinable horizon of openness to otherness, justice and political renewal. The violence used to defend and promote a political organization will always be required as the obligation of any system to its own continuity. Yet, it will also always draw on tropes of the greater good, of antagonism towards privileged or sectional interests and on conceptualisations of 'the majority,' or 'the people,' that will always evoke the imagined horizon of democracy-to-come as a license to an enthusiasm of the demos, its exaltation, its imaginary momentum and unimpeachable authenticity, and finally, its necessary historical realisation. In other words, the will of the people will always need to have its path opened by an intolerance to the (perhaps sometimes purely imaginary) partial, incomplete or sectional. Democracy is policed by illiberal violence, therefore, a violence that alone makes the operation of democracy possible. We can see this in the historical institution of democratic systems, whose violence is routinely romanticised as the midwife of freedom, or as necessary spasms of the human spirit. Democracy-to-come may be the term for the necessary orientation of human societies towards openness and justice but facing thus to the unimaginable also risks exposure to the open-ended messianic violence that any uninhibited call to justice may inevitably involve.

This violence may be deemed necessary in any particular instance, but it is violence nonetheless, a violence which is never acknowledged and is in fact automatically denied in the self-image of democracy. Over and above this, the inclusion of violence as a necessary, if disavowed, part of the practice of democracy unleashes a force not easily controlled. As we have seen in our account of Hobbes above, the institution of sovereign authority does not exclude violence in any simple way. It is, of course, violent itself, but more importantly, it also sews violence, what Hobbes understood as war, into the very lining of the social. The situation with democracy is similar. The call to democracy demands that its way be opened by an attack on the sectional, the aristocratic and esoteric, in the name of the human, universal and majority. These latter concepts are inexhaustible, and can easily become signifiers for unquantifiable demographics: the people, the silent majority, the average or real citizenry. In turn these categories can be represented as perpetually threatened, undefended and vulnerable, as being subject to the oppressions of new cabals, sometimes seen to be controlled—incredibly—by the most marginal and oppressed human groups. Instituting revolutionary, *messianic*, violence, to use Benjamin's term, is always available to democracy, as war is always threatening in Hobbes's sovereign order. It is easy to enlist this violence in wars fought in democracy's name.

The situation with human rights is analogous. The canonical statements of human rights—the *Déclaration des droits de l'Homme et du citoyen*, and the *Universal Declaration of Human Rights*, for example—were, of course, precipitated by revolutionary and global war, and arose either in order to simplify the prosecution of war or recover from it some meaningful sense of human progress. The world systems that offer to institute rights only gained their opportunity by way of war. In other words, the realisation of human rights is only possible in a world that war has created, a world that would not have been possible without war. As in Kant's argument, 'peace' only arises historically after war, and literally as its consequence. Peace is only possible by way of, and after, war. Human rights emerge by way of a founding messianic violence which they simultaneously require, on the one hand, and anathematise and spurn, on the other. The path to human rights is made possible by violence, and violence continually accompanies them, as the facilitation they may reluctantly need, but which they still hope to annihilate.

Derrida's 'democracy-to-come,' like justice in his earlier argument about law, is not ultimately realisable, but remains the irreducible and always excessive image of our aspiration. Every move we make to enact our sense of responsibility to the other's 'freedom,' our longing to 'let beings be,'[8] in Heidegger's phrase, represents our failed but dignifying tribute to this impossible democracy. Every actual democratic act, word and institution responds to it while being unable to live up to it. It is this open-ended commitment to an unachievable democracy that Derrida placed at the heart of what he called the New International, and his oft-repeated call for a New Enlightenment. Yet, we must remember that our open-ness to such an endlessly receding horizon of change also exposes us to risk. Messianic violence brings the danger of atrocity as much as the promise of renewal and re-founding. In an analogous argument, Derrida connects our day-to-day acts of hospitality to a general, open and undiscriminating absolute hospitality, one in which we welcome the stranger without even needing to know their name. Each of our mundane acts of hospitality is a homage to this automatic and inexhaustible hospitality, in which our first impulse is to honour, shelter and entertain the other without hesitation or question. *If only we could always be so generous!* Yet, Derrida complicates this wild hospitability with the example of Lot, whose unhesitating hospitality towards the angels—he does not even ask them to identify themselves—ends in the most appalling and cruel indifference when he offers up the least powerful members of his household, his young daughters, to be raped by the people of Sodom in the angels' stead.[9]

In sum, our necessary commitment to the absolute that promises our goodness also exposes us to what Derrida calls 'the menace of radical evil,'[10] or even 'the worst.'[11] Democracy-to-come is a wonderful promise but also a great risk. This risk is not something hovering purely in the domain of the possible or aleatory. It emerges constantly in the authoritarianism and violence used to institute or protect democracy. Not only is violence a bridge we might have to cross in order to reach democracy, but it is part of the appeal of democracy, its emphatic rightness that we seek to see everywhere acknowledged, its exaltation of the energy and goodness of 'the people.' The wars we fight and the ones we long to fight for it—from the revolutionary campaigns of the 18th century to the removal of authoritarian, racist and gangster regimes in our own time—are an implicit part of democracy's meaning, and its heroism. When we commit to the 'rule of law,' we should know from the phrase itself that we are committing to the violence that will have to be employed to ensure that rule. When we commit to human rights, civic freedom, representative government, executive accountability and a transparent judicial system, we are also committing to the violence which will not only protect them, but which enacts our faith in their necessity. Democracy has become obligatory in our age, and the fact that the human celebration of it has become compulsory means that it will always be accompanied by the violence which most discredits and threatens it. Despite our fantasies, this violence is not something to which we must resign ourselves as the sober custodians of a preferable order, one for which there is simply a cost. It is a program that can lead to the rationalisation of atrocity, the routinisation of collateral damage, the bureaucratisation of terror and the legitimisation of torture. The combination of the fear of violence with its volatile satisfactions, provides democratic societies with their most telling fault-lines: between the cosmopolitan and the patriot, radical pluralism and radical populism, those to whom democracy is a promise of absolute liberality and those to whom it is a promise of collective triumph, the new new left and the new new right, those to whom democracy has never gone far enough, and those for whom it has always already gone too far. In democracy's case, the absolute and unconditional quality of our commitments means reform and enthusiasm, the will-to-freedom and the will-to-dominance, progress and overcoming can never quite be kept apart. No wonder that democracy—even in countries where it has been instituted for centuries—still evokes a simultaneous excitement and bitterness: a social peace that is both a great relief and a numbing tedium, an horizon of change that is both a great adventure and a great danger. Democracy is our future but not our redemption. It must remain our problem.

Notes

[1] I. Kant, *Perpetual Peace*, L.W. Beck (ed), Liberal Arts Press, Indianapolis, 1957, p. 46.
[2] Ibid., p. 22.
[3] Ibid., p. 24.
[4] C. von Clausewitz, *On War*, trans. J.J. Graham, Penguin Books, London, 1968, p. 119.
[5] E. Levinas, *Totality and Infinity*, trans. A. Lingis, Duquesne University Press, Pittsburgh, 1969, p. 222.
[6] Ibid.

[7] B. Brecher, 'Why There is no such Thing as Political Terrorism,' *Metamorphosis of War*, A. Plaw (ed), Rodopi, Amsterdam and New York, NY, 2012, pp. 41-50.
[8] M. Heidegger, *Basic Writings*, D.F. Krell (ed), Routledge, London, 1993, p. 125.
[9] J. Derrida and A. Dufourmantelle, *Of Hospitality*, trans. R. Bowlby, Stanford University Press, Stanford, 2000, p. 153.
[10] J. Derrida, 'Faith and Knowledge,' *Acts of Religion*, G. Anidjar (ed), Routledge, New York and London, 2002, p. 83.
[11] J. Derrida 'Force of Law,' *Acts of Religion*, G. Anidjar (ed), Routledge, New York and London, 2002, p. 257.

Bibliography

Badiou, A., *Polemics*. Trans. Corcoran S., Verso, London, 2006.

Bataille, G., *Theory of Religion.* Trans. Hurley, R., Zone Books, New York, 1989.

Benjamin, W., 'Critique of Violence,' *Reflections: Essays, Aphorisms, Autobiographical Writings*. P. Demetz (ed), Schocken Books, New York, 1986.

Brecher, B., 'Why There is no such Thing as Political Terrorism.' Metamorphosis of WarDeleuze, G. and Guattari, F., *A Thousand Plateaus: Capitalism and Schizophrenia, Volume 2*. Trans. B. Massumi, B., University of Minnesota Press, Minneapolis, 1987.

Derrida, J., *Politics of Friendship.* Trans. Collins, G., Verso, London, 1997.

—, *Adieu to Emmanuel Levinas*. Trans. P.A. Brault and M. Naas, Stanford University Press, Stanford, 1999.

—, 'Faith and Knowledge'. *Acts of Religion*. Anidjar, G. (ed), Routledge, New York and London, 2002.

—, 'Force of Law,' in *Acts of Religion*. G. Anidjar (ed.). Routledge, New York and London, 2002, pp. 228–98.

— and Dufourmantelle A., *Of Hospitality*. Trans. Bowlby, R., Stanford University Press, Stanford, 2000.

Foucault, M., *Society Must be Defended.* Trans. Macey, D., Allen Lane, London, 2003.

Freud, S., *Civilization, Society and Religion.* Trans. Strachey, J., Richards, A. (ed), Penguin Books, London, 1985.

Hardt, M. and Negri, A., *Multitude: War and Democracy in the Age of Empire.* Penguin Books, London, 2004.

Heidegger, M., *Basic Writings.* Krell, D.F. (ed), Routledge, London, 1993.

Hobbes, T., *Leviathan.* Gaskin, J.C.A. (ed), Oxford University Press, Oxford, 1996.

Kant, I., *Perpetual Peace.* Beck, L.W. (ed), Liberal Arts Press, Indianapolis, 1957.

Levinas, E., *Totality and Infinity: An Essay on Exteriority.* Trans. Lingis, A., Duquesne University Press, Pittsburgh, 1969.

Schmitt, C., *The Concept of the Political.* Trans. Schwab, G., University of Chicago Press, Chicago, 1996.

von Clausewitz, C., *On War.* Trans. Graham, J.J., Penguin Books, London, 1968.

Nick Mansfield teaches Critical and Cultural Studies at Macquarie University in Sydney. His most recent book is *The God Who Deconstructs Himself: Subjectivity and Sovereignty between Freud, Bataille and Derrida*, (2010). He is one of the founding editors of the journal *Derrida Today.*

Foucault and the Continuation of War

Jason Edwards

Abstract
In this chapter, I'll focus on Michel Foucault's analysis of the relationship between war, power relations and politics in his series of lectures published as *Society Must Be Defended.* I'll argue that Foucault gives us good reasons to contest the liberal conception of politics as the suppression of war and the Clausewitzian view of war as a small-arm of politics. But I'll argue against the view that Foucault is providing us with a general account or philosophy of politics as the continuation of war. Some contemporary authors, following Carl Schmitt, tend to see the political as a realm of antagonism that adopts the form of war. I'll claim that such a view is of limited use and does not help us, as reading Foucault can, in understanding how politics and relations of power in modernity are constituted as relations of antagonism, or, in other words, how the construction of the friend-enemy distinction is characteristic of the political discourses and practices of modernity.

Key Words: Foucault, war, politics, the political, antagonism, friend-enemy distinction.

*****

1. Introduction

For a long time, the analysis of war was connected to the study of politics only in a certain Clausewitzian sense. War was what states engaged in to protect and advance their interests when diplomacy failed. In this respect, war was conceived of as an arm of politics, and one that tended to be employed only in exceptional circumstances. War, in this sense, is subordinate to politics. But what would it mean to turn this formulation on its head? What if we studied war not as the continuation of politics, but politics as the continuation of war?

In this chapter, I want to pursue this question by reference to Michel Foucault's series of lectures that are published under the title *Society Must Be Defended.*[1] I won't really be offering an interpretation of Foucault's arguments, but rather just addressing some of the problems that arise from considering politics as the continuation of war in the way that Foucault indicates. More specifically, I want to point to some of the difficulties with understanding politics as war that emerge from the assumption that there is no primal state of war and that war itself is always the product of a complex interplay of social discourses and practices. What conceptual and theoretical problems arise when we ask how social antagonism is constituted through relations of power, and to what extent it involves forms of force and domination? Is politics, in any meaningful sense, the continuation of war, or is this simply a way of talking in light of the kind of people we think and say we are in the world today?

The chapter is divided into three main sections. The first provides an overview of Foucault's arguments in *Society Must Be Defended.* The next section deals with the immediate problems that emerge from his use of war as a means of analysing politics and power relations. The third section focuses on some of the problems of using the concept of politics as the continuation of war today, in particular as that relates to the understanding of the political as a realm of ineradicable antagonism, and the friend-enemy distinction.

2. War Pursued by Other Means

In *Society Must Be Defended*, Foucault sets out to consider the question of the extent to which we can use war as an analyser of power relations, or, in his more general formulation, whether we can reverse Clausewitz's famous maxim declaring that 'politics is the continuation of war by other means.' The impetus for this inquiry is Foucault's rejection of the 'juridical model of sovereignty.' By this, he means a model of the state and politics which emerges in the sixteenth century, and which involves three basic terms—subject, law and sovereign. This model orients itself around the question of the legitimacy of sovereign power. In this model, sovereignty is the legitimate and sole source of all power in a society. Power, in this regard, is public power, and its nature is essentially negative. The sovereign has power over life and death, but, as Foucault claims, the juridical model effectively concedes that the real power that the sovereign possesses is over death. The sovereign can let live, that is, allow people to go on living, but letting live is always a passive act of omission. Sovereignty only really manifests itself when it kills—at least figuratively speaking. So, you might never be aware of this repressive power of the sovereign until it comes knocking on (or knocking down) your front door in the middle of the night. These acts of the sovereign are mandated in law. When this model of sovereignty arose in the sixteenth century, the operation of the law was seen as the consequence of the tacit or express agreement of subjects to be subjugated, that is to surrender themselves to the law that is an emanation of the sovereign's will. Power, and politics, then, come from the centre—that centre is sovereignty, and the form of power relations is encoded in the laws that carry the will of the sovereign. And from the sixteenth century, the right of sovereign power is seen not as a direct gift of divinity, nor in the status of the sovereign as *paterfamilias* in the line of Adam, but rather from this subjugation of and by the subjects to the will of the sovereign.

Now, we may know this doctrine under other descriptions—the social contract theory, the liberal conception of sovereignty, etc. But whatever its particular designation, the important point is that politics is regarded as the normal practice of a society in which sovereignty is founded on subjugation of subjects, and in which the legitimate sovereign power is exercised, on a legal basis, to kill (literally or figuratively) those who would challenge sovereignty, the law and the peace of the sovereign's subjects. Indeed, this juridical account of sovereignty lies at the heart of all modern constitutional and public law.

Politics, then, is conceived of as the very absence of war, and this lack of war is to be contrasted with the relations that operate between human beings who live in spatial proximity when there is no sovereignty—and hence no politics. In this situation, as philosophers like Grotius, Puffendorf and Hobbes have told us, war may not be ever-present, but the very possibility of war is indeed ever-present. Where there is no subjugation of subjects to a sovereign, there will be contending forces, contending groups, who persistently seek to dominate others. Relations of domination are therefore aleatory and subject to constant change in both scope and intensity. War breaks out because men are by nature more-or-less equal. No single set of relations of domination can endure purely on the basis of the exercise of violence by some or other individual or group of individuals. Such violence has no legitimacy in the eyes of the dominated, and thus they constantly seek to overturn these relations of domination and institute new ones of which they are the authors. Domination, of course (as Weber, most famously, tells us) exists under the exercise of sovereign power, but the relations of domination under the sovereign are legitimate. The violence that is exercised by the sovereign is recognised as being rightful in the eyes of the dominated. There is a relation of inequality between sovereign and subjects, but the effect of this is that with sovereignty, not just war but the very possibility of war, is eradicated. The violence exercised by the sovereign against individuals or groups of men is not just not legitimate war, it is not war at all, since war necessarily involves parties who are essentially equal (not, of course, equal in strength of arms, but equal in the sense of having a legitimate right to engage in war, to challenge and uphold relations of domination that are indeed sustained by virtue of arms alone).

There are several respects in which Foucault wishes to challenge the juridical doctrine of the state and sovereignty, and the picture it paints of the relationship between war and politics. First, his studies of madness, medicine, punishment and incarceration in modernity had persuaded him that the notion that the conduct of individuals is only shaped by power at the level of the operation of the law that emanates from the sovereign, from the juridical process, is wrong. The 'mechanisms' that shape the behaviour of individuals in an asylum, in a hospital, in a penitentiary, in a classroom or on a parade-ground, are far removed from discourses of sovereignty and the law, which for all intents and purposes originate and remain oriented around the question of the legitimate extent of royal authority. Power operates in modern societies, then, at a 'capillary' level; relations of force and domination are to be discovered in a whole range of social institutions and practices that are structured only partially by the laws that come down from the heights of sovereignty.

Second, it follows that what we might call 'subjectification' is a product of the operation of these local and diverse relations of force and domination that operate at the very periphery, and indeed sometimes beyond the peripheral vision, of sovereignty. Subjects are made by these micro-relations of power. We must pose to the juridical doctrine of sovereignty the question of where the subjects who are subjugated by sovereign power come from in the first place. Rather than taking this individual who is capable of alienating his rights, or recognising the legitimacy of sovereign right, we must acknowledge that 'the individual is one of power's first effects.'[2] Just as much as we must inquire into what social relations of force and domination turn a person into a clinically confined madman, a patient, a soldier, a student or a prisoner, we must also inquire into what relations of power transform a person into a subject of sovereign power, and a person who has the capacity to authorise political power.

Thirdly, then, Foucault wants to resist the Hobbesian model of sovereignty as the sole source of power, emanating from a sovereign centre, and in which the legitimacy of the sovereign has been established by the alienation of all social or public power from individuals who are already invested with the necessary powers of freedom to voluntarily subject themselves to subjugation.

For Foucault, then, we cannot understand power relations and politics in modern societies on the model of the juridical discourse of sovereignty. That discourse sets out to justify royal authority, the power of the sovereign, by effectively disguising the fact that such power has been established through relations of force and domination, and that it must be sustained over time through relations of force and domination. Power is compacted, so that it is seen as the medium through which the sovereign exercises his right over the life and death of his subjects. Yet this disguises the fact that relations of power operate in the hinterland of sovereignty, working within and across non-juristic forms of discourse and practice to create subjects at a micromechanical level. Accordingly, the juridical model of sovereignty cannot aid us in analysing power relations and politics in modern societies and we must turn to an alternative model. It is here that Foucault raises the question of to what extent we should embrace a model of politics as the continuation, rather than the repression, of war.

We can explode the fiction of sovereignty being the result and expression of a free surrender of men's natural powers to the sovereign by recognising power relations for what they are—relations of domination and force. Foucault is not, of course, the first to conceive of politics and power relations in such a way. But he presents two different ways in which this understanding of power may be articulated. First, power or force relations are essentially repressive in character; power represses 'nature, instincts, a class or individual.'[3] This is a view of power that we can find at work in Hegel, Freud and Reich (indeed, Foucault dubs it, 'Reich's hypothesis'). Second, power relations can be conceived of as in essence a warlike struggle between contending forces, a notion Foucault identifies with Nietzsche (so, 'Nietzsche's hypothesis'). But these two notions should not be held to be in conflict; rather, they can be seen as being complementary. Repression is something that is always established by the application of force in war, with the victors consequently repressing the conquered. If we take this war-repression scheme as the vehicle for the analysis of power relations, then we should cease to see the question of legitimacy as being at the centre of the study of politics, and rather focus on the question of victory. But much of Foucault's work prior to *Society Must Be Defended* involves a good deal of scepticism about the view of power as being essentially repressive (and this comes out most obviously in the first volume of *The History of Sexuality*, published shortly after he gave the lectures making up *Society Must Be Defended*). The concept of repression, as it appears for example in Reich's account of the repression of infantile sexuality, often relies on attributing to people or groups qualities that are natural, at least in the sense that they are not constituted by or can operate independently of relations of power. Liberation, on this view, is seen as the overcoming of power to permit the free expression of these natural qualities—hence, for Reich, the importance of sexual freedom as a vehicle of revolution against bourgeois society. But for Foucault, while we can certainly see to an extent in relations of power the mark of a will to domination, we need to see how that which is being 'repressed' is itself constructed through the operation of power relations. Reich's 'infantile sexuality,' in other words, is not something that is natural or primordial, but rather is constituted through a complex set of scientific, technical and ethical discourses surrounding sexuality that are articulated in nineteenth century psychiatry and later in psychoanalysis. Generally speaking, all subjects that are subject to repression must first be constituted as subjects of repression by relations of power, and more specifically, the way in which these relations are manifested in particular discourses, social institutions and practices.

If, then, we can connect together Reich's and Nietzsche's hypotheses, it must be in the sense that we can never allow either for a static, permanent regime of repression, nor for the overcoming of repression. What is repressed has to be constructed by relations of power, and as we have seen, Foucault regards such relations as multiple, often contradictory and aleatory. At the same time, the condition of achieving liberation from repression, in the

Reichian sense, is to secure a way of life centred on ethics and practices that have been freed from the operation of power. But it is clear that Foucault believes such a notion to be fantastic. Ethics and practices, and the way in which these interact, require the existence of subjects who are constituted, and must always be constituted, through social relations of power. With respect to the analysis of politics by the repression-war scheme, then, if we see politics as the waging of a war that lies behind the operation of apparently peaceful processes, that war can never have a conclusion; there can never be peace. There will, at different times, always be victors and vanquished, but power operates in such a way that the vanquished can become the victors. The struggle for dominance and against submission constantly goes on in the very fabric of power relations, and the discourses and practices that it works through. It is this perpetual struggle and contest in modern societies that Foucault claims can be analysed in terms of relations of force, of strategy, and of tactics. We can indeed understand politics as the continuation of war.

The bulk of *Society Must Be Defended* is concerned with the question of where the notion of understanding politics in terms of war comes from. Foucault's aim, then, is not just to advocate the model of war as a means of analysing politics, but of tracing how that idea is constructed in the discourses of politics in the West from the sixteenth century. And indeed, there is yet a further goal of the lectures, which is to consider how this discourse of politics as war comes to be transformed into a discourse of state racism, a discourse that focuses not on a struggle of 'races,' understood in the broad sense of groups of men with a united national or cultural inheritance, but on a struggle between society as a whole and the biological threats posed to it—of degeneracy, counter-races and sub-races.

The discourse of war as 'race' conflict emerges in popular and aristocratic struggles against absolutist monarchy in the seventeenth and eighteenth centuries. It focuses on how the society created by absolute monarchs and their juridical apparatuses are not, as the defenders of absolutism would have us believe, the outcome of the free surrendering of men's powers to the sovereign in return for the protection of social peace for all, but rather the product of, and continuing guarantor of, the king's victory over his enemies. This victory is not hypothetical, in the way that the social contract is, but is a real historical event. In England during the 1640s, writers like the Levellers would portray this historical event as the Norman Conquest and the subsequent 'Norman yoke' that had trampled over the original rights and liberties of Anglo-Saxon men. The institutions of the Norman monarchs and their successors, therefore, in fact represented the continued waging of war by the apparatuses of the monarchy and its aristocratic allies against 'free-born Englishmen.' In France, aristocratic writers in the early eighteenth century, such as Boulainvilliers, saw the French monarchy as having its foundations in the Franks invasion of Gaul in the fifth century. Foucault claims these discourses to be new in the sense that they are historico-political, and oppose themselves quite explicitly to the counter-discourse of the jurdico-philosophical conception of the state. An understanding of history, as the field of warfare, is the way in which the abstract justifications of monarchical power by its jurists can be contested. In contesting them, the advocates of the historical approach were engaging in political acts, a quite overt struggle over knowledge to be wielded in a fight against their adversaries. The 'truth' of history, then, is something that can only be gleaned historically, and not by recourse to abstract theorising. War runs through history, and history is the history of wars and battles, not of peace.

By the time that Clausewitz writes that 'war is policy pursued by other means,' the reverse of this claim has been held to be true for at least a couple of centuries. The discourse of politics as war is, therefore, nothing new. But it is transformed in the Enlightenment, when it becomes, in the hands of the representatives of the Third Estate writing against the monarchy, a combat that takes place between the nation and those elements that are opposed to the nation. For Emmanuel Sieyes in his famous, 'What is the Third Estate?,' his answer to that question is that it is everything, that it is the nation, or as Foucault shows, that it is what makes the nation possible. The nation, then, becomes universal, and, moreover, to be identified with the state. In opposing the Revolution, Louis XVI and his clerical and aristocratic allies effectively excluded themselves from the universal nation, and appeared as enemies of the nation and the state. The Revolution represents a war in which the right of the nation is re-established, but for Sieyes this 'nation' is not now just one order of society, not, as it had been for the reactionary aristocrats, one nation existing within a territory with several nations, but the nation is everyone. Nationality is denied to those who oppose the Revolution—they are simply not recognised as part of the state that expresses the universal will of the nation. What we see here, of course, is the articulation of the modern notion of the nation. It is a notion that draws on the ideas of race struggle that Foucault outlines in his lectures as being constitutive of a historico-political discourse that opposes itself to the legitimising, juridical discourse of royal power or sovereignty. However, at the end of the eighteenth century, the bourgeois or liberal understanding of this historico-political discourse is that it allows us to recognise the point at which national universality brings about a dissolution of all war. War ceases with the casting off of absolutism, and with it goes the relations of domination that had marked all hitherto history.

What occurs in the French Revolution, then, is that 'politics,' where it is conceived of, no longer appears as war. Politics is the function of a state that is the representative of the unity constituted by a single nation. War is no longer possible because the combatants have been disarmed, and, in an important sense, the end of the *ancien régime* marks the termination of that struggle between different forces for domination that had characterised the history of the West. What we see here is a liberal discourse of politics—or, it might even be said, anti-politics—that is at work in the writings of Hume, Smith, Kant, Constant, Bentham and Mill. It recognises the fiction that is the social contract and the juridical account of sovereignty, and instead asks us to analyse politics in terms of historical struggle and change. At the same time, however, it sees in the Third Estate, commercial society, or the bourgeoisie, the announcement of the end of war. The 'politics' that goes on in modern societies performs or should perform a mainly administrative function, as part of an economy that is structured by the needs of the nation or the people. Politics in the modern world is thus not war, but war overcome. What we have here, then, is not the reinstatement or rearticulation of the juridical doctrine of sovereignty (though this has by no means disappeared), but rather the use of historico-political discourse to affirm that the modern state represents the universality of all, of the nation, and works in the national interest. We arrive, then, at the liberal conceptualisation of politics as the very opposite of war.

Yet Foucault is insistent that the discourse of politics and history constituted by war does not disappear in the wake of the French Revolution. Rather, the theme of 'race' struggle comes to be worked into a 'state racism.' This state racism is linked to the emergence of a new 'technology' of power in the eighteenth century that fundamentally differs from the 'disciplinary power' that is the focus of Foucault's famous book *Discipline and Punish*. Whereas disciplinary power had as its object the body of the individual and was concerned with its training and disciplining through everyday routines in the prison, the clinic, the school and the factory, this new form of power—what Foucault calls 'biopolitics'—focuses on the human race, or populations, as a whole. We move, then, from a concern with the regulation of individual conduct to a concern with the health of the race or population. The knowledge that is constructed by this biopolitical power is expressed in the science of population and health—the statistics of birth-rate, birth control and morbidity. And in turn, this knowledge functions through the institutions and practices of medicine and public hygiene. Biopower operates then, as the name suggests, at both a scientific and political level—it is a political discourse of biology.

Biopolitics is effectively a politics of life. The theories of sovereignty that emerged in the sixteenth century attributed to the sovereign the power over life and death, but we have already said that in effect this power was negative—it was in the main manifested in the sovereign's action of killing (or, more broadly speaking, coercing and punishing). It was the right of the sovereign to kill and to let live. But with biopower, there appears a form of power that 'consists in making live and letting die'—what Foucault calls, 'the power of regularisation.'[4] This regularising power is concerned with promoting the birth rate, lowering the mortality rate, raising life expectancy, and so on. It does this through instruments that are not disciplinary in character, not focused on training and disciplining individual bodies, but through mechanisms that 'establish an equilibrium, maintain an average, establish a sort of homeostasis, and compensate for variations within this general population and its aleatory field.'[5]

But surely there is a paradox here, for how can we see a discourse that is effectively concerned with preserving and extending life as informing 'state racism,' which is exactly about purging the nation of degenerate racial characteristics, and which, in the form of colonialism and imperialism in the late nineteenth century, is about suppressing and killing other races and civilisations? For Foucault, this is because in the late nineteenth century, a discourse of evolutionism (not, it should be stressed, Darwinism[6]), comes to see history in terms of a struggle for survival between races, with the fittest claiming the ultimate victory. It was in this way that a concern for the political regulation of the biological health of the racial or national population comes to lie at the centre of state racism and a new race war. In this guise, the 'races' are defined not in terms of cultural or social characteristics, as they are for the aristocratic and bourgeois opponents of absolutism in the seventeenth and eighteenth centuries, but rather by their biological attributes measured in comparison to those of other races. It is this discourse, combined with an account of how war is necessary both to protect one's population from degenerate elements, but also to expose it to the kind of aggression and violence that will root out its weakest elements and leave the population as a whole fitter, that gives rise to the kind of 'state racism' most obviously instantiated in the Nazi regime. Foucault concludes his analysis in *Society Must Be Defended* with an account of the Nazi conception of power as racial struggle that is articulated both through the mechanisms of biopower (Nazi race science) and an affirmation of the absolute right of the sovereign to kill the enemies of the race—even when those enemies are members of the race.

3. Contesting Politics as War

In this section, I want to consider some of the theoretical problems that arise from Foucault's assertion that politics can be analysed as war pursued by other means. But in doing so, there are two lines of thought that I want to avoid. First, I don't want to embark on an analysis that is about showing what Foucault 'really' meant. As in much of his research and writing, what Foucault was concerned with in his lectures on war and politics was posing new questions and opening up new lines of enquiry. He was not really interested in providing definitive accounts of explanatory concepts, or constructing a master-key for the analysis of social relations. In this respect, it's not clear that Foucault is trying to provide a definition of *the* concept of politics, rather than posing questions about what we might understand politics to be in a simultaneously historical and theoretical sense. The purpose of that understanding is, ultimately, to achieve a better comprehension of the kind of world we live in and the kind of people we are, and by so doing to realise something about how social life can be changed. Second, and strongly related to his point, I don't myself want to try and provide a definition of *the* concept of politics or of 'the political,' for the reason that there are many different concepts of politics and of the political. The criteria for their evaluation is not a real object called 'politics' or 'the political' that exists outside of political discourses and practices, but rather the role that they perform in furthering our understanding of how these discourses and practices operate.

There are, then, at least two senses in *Society Must Be Defended* in which we can understand the notion that politics is the continuation of war. First, we might consider that war is just the central feature of human history and politics, that what most of history is about is the waging of war and the manner in which the victors in war continue to impose their triumph on the vanquished. If we want to understand where we are and what it means to live in the modern world, we have to see ourselves as occupying positions in an ongoing and perhaps immemorial war. That idea, as we have seen, was for Foucault at the heart of what he describes as the 'historico-political' discourse of politics that emerged in the seventeenth and eighteenth centuries as a response to absolute royal power. This discourse of politics as the continuation of war sees history as a constant series of battles—that, if you like, constitute a seamless war—in which the original conquest is never complete, and in which victory is always fragile and open to reversal by the defeated. In modernity, the jurdico-philosophical discourse of sovereignty has sought to disguise this on-going war by seeing the source of sovereignty as lying in its authorisation by free men, and the replacement of force by law.

This discourse of politics as the continuation of war was initiated by the aristocratic and popular opponents of absolutism. But in important respects, many of the most prominent twentieth-century theorists of the relationship between politics and war have adopted some of the central tenets of this historico-political discourse—for example, Max Weber, Carl Schmitt and, more recently, historical sociologists of the state like Charles Tilly. States are seen as having their origins in organised violence, conquest and the subjugation of internal enemies by a victorious monarchy. There is a sense, then, in which the discourse of politics as war, which has its origins in active political opposition to the fledgling absolutist state, might today be employed as an analysis of state-society relations. But we also need to recognise, as Foucault does, that it is precisely this discourse that gives rise to a conception of 'race struggle'—that the war that underlies political and legal institutions and practices is one embedded in irreconcilable conflict between enemies, enemies defined in terms of their essential differences, whether those be differences of biological race, ethnicity, class or nationality. And in that sense—and indeed in the work of many twentieth century sociologists and historians of modern state-society and international relations—this discourse sees these categories of identity as existing prior to and as a precondition of war and politics. This is an idea that, along with Foucault, we must contest—such identities can themselves be seen as a product of war and politics.

The identity of the combatants in war is, of course, historically contingent. Today's enemies may be tomorrow's allies, and our allies in today's victory may be our enemies tomorrow. The Reformation and the Wars of Religion in Europe carved out a series of shifting identities that had a central role in the formation of the modern state and state system. The reasons for the emergence of a Protestant movement are complex, but, without questioning the fact that the 'politics' of the Roman Church in the sixteenth century had much to do with it, the religious schism in Europe did not map neatly onto existing political and territorial enmities. The power of Protestant ideology to unite and divide (sometimes within the same family) reshaped political organisation and loyalties. At the closing of the Wars of Religion, as Hobbes clearly understood, fatigue with war, both within and between states, had compelled religious antagonists into the acceptance of the right of the sovereign to determine the confession of his subjects—the doctrine of *cujus regio ejus religio* established in the Peace of Westphalia. Westphalia can be seen as representing the triumph of politics—or at least the conception of politics as the sovereign's legitimate protection of civil order and peace—over war. The subjection of subjects to the sovereign was, in this sense, a real phenomenon, and not simply a justification of royal authority by theorists of sovereignty like Hobbes.

There are, then, good grounds on which we should distrust the notion that politics is the continuation of war if that is taken to mean that politics is simply war in disguise between combatants whose identity is pre-political, or that it is only through war that, historically speaking, we have arrived at the forms of political organisation and identities characteristic of modernity. It is true that for much of the twentieth century, war and warfare were often overlooked in political and social thought, but that is no reason, from the point of view of the analysis of state-society and international relations, to posit the notion of the primacy of war and warfare in the constitution of political modernity. While we do indeed need to bring the war-politics distinction into question, the complete erasure of that distinction distorts the complex character and aetiology of the modern state, its relation to society and the nature of the international system.

The second sense in which we might understand the notion of politics as the continuation of war is in the guise of what Foucault described as 'Nietzsche's hypothesis.' Here, the lines between the historical and the philosophical become blurred as we conceive of all power relations as, in essence, a warlike and unending struggle between contending forces. In this regard, we have to shift our understanding of politics from the circumscribed realm of the state and its laws, and the relation between state, society and the international system, to all social relations of power. These relations operate in the hinterland of sovereignty, working within and across non-juristic forms of discourse and practice to create subjects of power at the micro-mechanical level. This is the understanding of power relations that is at work in Foucault's ideas of disciplinary power and biopolitics.

Foucault has been accused by his critics of uncritically embracing a Nietzschean metaphysics of power, and as a result, by logical inference, he should be found guilty of a kind of ethical and political nihilism. But, for reasons that I can't really elaborate on here, I don't think this is a compelling criticism of Foucault's understanding of power relations.[7] Rather, I think the clear attraction Nietzsche had for Foucault was that he showed that identity and resistance to domination and subjugation are formed *within* relations of power, and not on their outside by virtue of God, natural law or natural reason. The appeal for Foucault of the historico-political discourse of race struggle is that it revealed the myth of sovereignty. The sovereign state does not eliminate the war-like struggle between contending parties that goes on in its territory. Rather, the state is drawn into these struggles, even at its periphery where its laws do not operate. This historico-political discourse provides us with a model for understanding power relations as having the form of war, even if the state does not recognise this, even if it is philosophically justified in terms of rightful consent and its provision of protection in return for obedience. Sovereignty functions to disguise the war that constantly goes on around the state, and which unavoidably draws it in.

On this second understanding of politics as the continuation of war, then, power relations are war-like in character. They involve struggle, and tactics and strategies for domination and subjugation of the enemy. This is not a war of the state on society, or even between states, it is a war that goes on within both the state and society. We're likely to be averse to this kind of understanding of all social relations of power as the continuation of war because of the way in which we usually talk and think about war. For we often think that war should be defined as a means of action and an instrument of conduct (as, in other words, organised violence oriented towards a rational end, involving specially trained and armed forces, involving rules of combat, etc.), rather than as a *relationship* between contending parties. However, we also speak of war existing even where there is no violence, because there's a clear sense that what counts is the war-like relationship. A war-like relationship exists wherever we can identify antagonists and where—crucially—the contending parties recognise themselves as antagonists. And I think this is the sense in which Foucault wanted to employ the concept of politics—or power relations—as the continuation of war, a concern that emerges in his considerations of power. For Foucault was keen to contest the notion of power as absolute domination. Indeed, power could not operate in the absence of resistance. Domination, in that sense, can only ever be a process of active subjection, of the constant repression of resistance. The subject of domination, on this view, is not a passive and defeated figure, but a resisting agent, using any resources available to it—including those provided by the dominator—in order to contest its status in the relationship.

From a certain point of view, then, we can understand war as analyser of politics in the sense that social relations are antagonistic in character, and that the constitution of the relationship between the antagonists is a dynamic and contingent process that is never closed or completed. It might be, then, that politics is indeed like war in a certain Clausewitzian sense. *On War* paints a picture of war and its outcome as dependent on the relation between the warring parties, characterised by the visceral impulse to defeat the enemy, the random play of contingent factors and the rational calculation of means and ends.[8] Clausewitz was writing about war in a particular context and to a particular end. The Napoleonic Wars had brought about a revolution in the organisation and conduct of military campaigns, had mobilised the masses in nations in the waging of war and had introduced the spectre of 'total war.' But Clausewitz's work is often taken as not simply providing a commentary on the state of war in the modern world, but rather outlining a general philosophy of war. It is in this sense that the form that war takes on through human history—the interplay of the triad of visceral emotion, contingency and reason—might be thought to be at the same time the form of general social relations. The difficulty with drawing such an analogy appears to be Clausewitz's insistence that war is always the application of instruments of violence to force the enemy into submission. However, what constitutes 'violence,' 'arms' and 'force' in this regard is extremely problematic. If we are talking about a general form of war that may go on in society without the kind of formal hostilities that we have become used to in the age of conflict between the trained military forces of states in the international system, then we may think of many types of instruments of violence, force, coercion, that are employed in the ordering of social relations. It is, therefore, no bar to a general conception of relations of power as relations of war if there is an absence of specifically martial institutions and practices.

Prior to his lectures in *Society Must Be Defended*, Foucault had explored—via his conception of disciplinary power outlined in *Discipline and Punish*—the manner in which many social relations and institutions in modernity came to be organised by adopting the same mechanisms of control and surveillance that operated in the training and deployment of soldiers. But what he develops in the lectures in *Society Must be Defended* departs from this kind of analysis by focusing on the *form* of war as the form of power relations. It's in that sense, then, that we might think of Foucault adopting something of Clausewitz's philosophy of war. Soldiers, like pupils, workers and the mentally ill, are all made through relations of power. But these power relations, like relations of war, are antagonistic in character, and their outcome is always uncertain. In the final section of the chapter, I'd like to turn towards the status of this claim. What precisely, does it mean to say that we can think of power relations as having the form of war by virtue of the presence of an antagonistic relation between the parties involved? And can we then define politics—or the political—as being in essence defined by the relation of antagonism?

4. The Political, War and Antagonism

Carl Schmitt's *The Concept of the Political* makes a famous claim about the character of 'the political.'[9] Schmitt presents the political not as a specific institution or set of practices—that which we usually identify with 'politics'—but as that which makes possible the distinction between friend and enemy. It would be a mistake to regard Schmitt as saying that the political is war, but the political is certainly what makes war possible. War is not simply the playing out of rivalries, competitions or personal hatreds, but rather a conflict that takes place between enemies. The enemy is *hostis* not *inimicus*—i.e., it is the enemy that threatens public safety, that which is entirely alien or opposite to who we are. In that sense, the political is what determines the identity of any group, because such an identity can only be established when 'we' have knowledge of who we are not—that which is our opposite and which (potentially at least) threatens our being by its own being.

From this point of view—and as contemporary authors such as Chantal Mouffe have argued—we can see the political as a relation of enmity or antagonism. Mouffe takes Schmitt's work to demonstrate that the political 'must be conceived as a dimension that is inherent to every human society and that determines our very ontological condition.'[10] In that respect, we might understand Mouffe as arguing that the political is not a departure of or cessation of war, but rather that the political has the form of war and is the condition for and moment of the constitution of all social life. It is not then, from this point of view, a question of war going on beneath the surface of politics and everyday life. Rather, politics is the continuation of war in the sense that the very social and political identities that form the parameters of all human social conduct are made possible by the constitutive relation of antagonism that is the political. The relation that is definitive of war—the distinction between friend-enemy—is the relation that is definitive of the political.

Mouffe's understanding of the political is designed to counter liberal conceptions of politics as the suppression or elimination of antagonism. In doing this, however, she presents us with a general metaphysics of social life as constituted by antagonism. If we consider identity as relational, then it is indeed the case that antagonism plays a central role in its formation: we can only comprehend what we are through the speculum of the other. It is, however, something of a leap from this proposition to the idea that the political should be conceived of as the friend-enemy relation and that antagonism is the ground of social life. Schmitt's work can indeed be read as an important corrective to the liberal view of politics and the law as the elimination of war and social antagonism. But the notion that the political is the continuation of war because both are at bottom a relation of antagonism, only serves to raise more questions than it answers. One such question is why we should accept this conception of the political in the face of alternative republican accounts, such as that of Hannah Arendt, that stress the difference of the political as the realm of freedom in which human beings act in concert to achieve the public good, to set out and obtain collective goals? In fact, we search in vain for any Archimedean point from which to judge the truth of such philosophical claims about the nature of the political. We may understand the concept of the political either as the condition of the antagonistic constitution of social identity or the condition of public action and collective endeavour. Whether we understand them as such depends on their use and place in any given political discourse.

The critique of a political or social theory can all too easily fall into the construction of a philosophy of the object under consideration that fails to reflect on its own political and social conditions of articulation. I'd like to suggest that Foucault's analysis of war raises a series of conceptual problems that are not easily answered, but at the same time the kind of analysis he performs allows us to bring into the question the very posing of the philosophical question of the nature of war and of the political. Foucault himself never engaged directly with Schmitt's work, but we can assume that he would have had some sympathy with Jacques Derrida's deconstruction of the friend-enemy distinction as outlined in Nick Mansfield's piece in this volume.[11] Derrida argues that Schmitt's friend-enemy distinction is reliant on the existence of the public-private distinction (since the enemy is *hostis*—the *public* enemy who may at the same time be a *private* friend) that can only be brought into being by the friend-enemy relation. Schmitt's failure here is related to his inability to recognise the problem that lies at the heart of Hobbes's account of the constitution of sovereign power. For as much as Schmitt looked to Hobbes as legitimating the protection-obedience relation, like Hobbes he takes it for granted that men are already endowed with the capacity to be the subjects of sovereign power. But for men to actually live under law, to accept the protection of the sovereign in return for obedience to the law, can only be possible once men recognise themselves as subjects of the law and of the sovereign. In turn, this requires a knowledge of themselves, of others and their actions.

Foucault, of course, recognises this problem as being embedded in the discourse of sovereignty, one in which Hobbes—and Schmitt—are mired. Schmitt famously rejected the legalistic account of sovereignty by claiming that sovereign power emanates not from the law but from the decision: 'sovereign is he who decides on the exception.'[12] But what is not brought into question on this view is the *capacity* of the sovereign to decide. Schmitt and other critics of the juridical account of sovereignty expose the fallacy of the liberal justification of sovereign power, but in the course of doing so they fail to ask the all-important question of from where the capacity to be sovereign, to make the decision about who constitutes 'friend' and 'enemy,' is derived. For Foucault, of course, the reality is that the King's head was cut off a long time ago, and that what it means to be a subject in modern societies is shaped by the plurality of powers that are liminal to the law and sovereignty. From this perspective, the friend-enemy distinction is not the metaphysical precondition of human social life, but is rather constructed in a variety of discourses and practices that are characteristic of—if not, perhaps confined to—modernity.

The politics-political and the friend-enemy distinction are, then, products of the kind of societies we inhabit in modernity, societies in which the 'public' is widely expected to participate in contesting 'politics.' As Reinhart Koselleck has shown, we may understand the 'Enlightenment' as the key moment in the formation of these contesting subjects.[13] The Enlightenment created a public that was equipped with the conceptual armoury and moral vocabulary to take on the authority of royal absolutism. But as Foucault indicates in *Society Must Be Defended*, the knowledge that the 'bourgeois' public creates involves not just the identification of the monarchy and the church as the enemy, but anything that is opposed to the nation. The Enlightenment and the French Revolution—and perhaps modernity in general—turn on this recognition of the national will and the defeat of its other, whether that other be an internal or external threat. Identifying the enemy, in this respect, means the creation and contestation of a whole new vocabulary of politics. There is a war over language that, of course, still goes on. Bob Brecher's chapter in this volume demonstrates one form of contestation of language in the present—the manner in which the term 'terrorism' is today employed to both delineate the enemy and to render it politically and morally bereft.[14]

Reading *Society Must Be Defended* highlights the many potential dangers in constructing a philosophy of antagonism, one which sees, for all intents and purposes, the form of war as being the form of politics or the political. That is not to say that we cannot understand political modernity as being, from one very important point of view, about the perpetual construction of the friend-enemy distinction that takes place in a very broad array of political discourses and practices. But an understanding of the world requires a demand to look at precisely how specific antagonisms have been constructed through the various concepts, vocabularies, discourses and practices of politics that we have inherited. It is such a project—rather than a philosophy of antagonism—that, I think, Foucault was trying to outline in *Society Must Be Defended*, and if we want to understand how the friend-enemy distinction plays out in the present, I would suggest that this project constitutes a good starting point.

5. Conclusion

How might it be a good starting point? First, and very crudely speaking, the international system can be analysed from the point of view of at least two discourses: one that sees it as the outcome of perpetual conflict between its constitutive, self-interested, units; and the other that sees the system as bound by customary, and increasingly formal, international law. We might, then, see in this a reflection of the distinction that Foucault makes in *Society Must Be Defended* between a historico-political and juridical account of the state and sovereignty. But what Foucault's analysis can bring into relief is the way in which the international system and the wars that occur within it, are not simply objects of one or the other of these discourses. Rather these discourses are themselves centrally implicated in the functioning of the system and the legitimation of warfare. 'Realist' and 'legalist' doctrines of international relations and war are positions taken up by contending parties in the crafting of the rules of the system and in the justification given for going to war.

Second, it's clear that an understanding of 'war' in the international system today as armed conflict that takes place between the specialist martial organisation of states under the laws of war is hardly adequate for comprehending the character of international conflict. Nor can we achieve an effective understanding of the international system by viewing it as the product of diplomatic relations between states. While it is important to recognise the continued importance of states and their 'high' politics as the major players in the international system, it's clear that a whole range of groups and conflicts below and across the level of states are central to the way in which global conflict is organised today—from micro-nationalist separatist conflicts, to Maoist guerrillas, to al-Qaeda. If we conceive of such groups and conflicts as operating 'beyond' politics, as being fleeting expressions of ancient ethnic rivalry, a commitment to atavistic ideology or fundamentalist religion, then we overlook the importance of liminal conflicts in structuring relations of power in everyday life.

Thirdly, and related to this last point, Foucault's analysis in *Society Must be Defended* draws our attention to the manner in which everyday 'politics' in modern societies operates around a constant vigilance and struggle against the enemy. There is no room here to enter into a discussion of Foucault's concepts of 'biopolitics' and 'governmentality.' However, the regulation of the conduct and behaviour of the entire population in order to preserve and extend life, to protect it from the internal and external enemies, is clearly a central feature of our modern lives. Whether it be the War on Drugs or the War on Terror, we need to be concerned about the profound impact of the enemy—an enemy that is produced and reproduced in the discourses and practices of politics in which we are immersed—on our everyday lives.

Notes

[1] Foucault delivered the lectures at the College de France in 1975-6, in between the publication of two of his most famous books: *Discipline and Punish* (1975), and *The History of Sexuality, Vol. 1* (1976).
[2] M. Foucault, *Society Must be Defended: Lectures at the Collège de France*, trans. D. Macey, Penguin Books, London, 2004, p. 30.
[3] Ibid., p. 15.
[4] Ibid., p. 247.
[5] Ibid., p. 246.
[6] Ibid., p. 256.
[7] This is only partly because the widespread view of Nietzsche as a nihilist is itself wrong. More importantly, Foucault's interest in Nietzsche, at least in the late 1970s, was clearly focused on how genealogy repudiates the search for the historical origins and metaphysical ground of truth. But this is far from saying that power is all that there is. See M. Foucault, 'Nietzsche, Genealogy, History,' *The Foucault Reader: An Introduction to Foucault's Thought*, P. Rabinow (ed), Penguin Books, London, 1984.
[8] C. von Clausewitz, *On War*, Penguin Books, London, 1982.
[9] C. Schmitt, *The Concept of the Political*, trans. G Schwab, University of Chicago Press, Chicago, 1996.
[10] C. Mouffe, *The Return of the Political*, Verso, London, 1993, p. 3.
[11] N. Mansfield, 'Fighting for Peace: From the Social War to Armed Democracy,' *The Metamorphosis of War*, A. Plaw (ed), Rodopi, Amsterdam and New York, NY, 2012, pp. 11-13.
[12] C. Schmitt, *Political Theology*, trans. T.B. Strong, The University of Chicago Press, Chicago, 2005.
[13] R. Koselleck, *Critique and Crisis: Enlightenment and the Pathogenesis of Modern Society*, The MIT Press, Cambridge, Mass., 1988; see also, J. Edwards, '*Critique and Crisis Today*: Koselleck, Enlightenment and the Concept of Politics,' *Contemporary Political Theory*, Vol. 5, 2006, pp. 428-446.

[14] B. Brecher, 'Why There is No Such Thing as Political Terrorism,' *The Metamorphosis of War*, A. Plaw (ed), Rodopi, Amsterdam and New York, NY, 2012, pp. 41-50.

Bibliography

Arendt, H., *The Human Condition*. The University of Chicago Press, Chicago, 1998.

Brecher, B., 'Why There is No Such Thing as Political Terrorism'. *The Metamorphosis of War*. Plaw, A. (ed), Rodopi, Amsterdam and New York, NY, 2012.

Edwards, J., 'Critique and Crisis Today: Koselleck, Enlightenment and the Concept of Politics'. *Contemporary Political Theory*. Vol. 5, 2006, pp. 428-446.

—, 'The Ideological Interpellation of Individuals as Combatants: An Encounter between Reinhart Koselleck and Michel Foucault'. *Journal of Political Ideologies*. Vol. 12, No. 1, 2007, pp. 49-66.

Foucault, M., *Society Must be Defended: Lectures at the College de France, 1975-1976*. Trans. Macey, D., Penguin Books, London, 2008.

—, *Discipline and Punish: The Birth of the Prison*. Trans. Sheridan, A., Penguin Books, London, 1991.

—, *The History of Sexuality, Volume 1: The Will to Knowledge*. Trans. Hurley, R., Penguin Books, London, 1998.

—, *The Birth of Biopolitics: Lectures at the College de France, 1978-1979*. Trans. Burchell, G., Palgrave MacMillan, London, 2008.

—, *Security, Territory, Population: Lectures at the College de France, 1976-1977*. Trans. Burchell, G., Palgrave MacMillan, London, 2008.

—, 'Nietzsche, Genealogy, History'. *The Foucault Reader: An Introduction to Foucault's Thought*. Rabinow, P. (ed), Penguin Books, London, 1984.

—, 'Governmentality'. *Essential Works of Foucault 1954-1984, Vol. 3: Power*. Faubion, J.D. (ed), Penguin Books, London 2002.

—, 'The Subject and Power'. *Essential Works of Foucault 1954-1984, Vol. 3: Power*. Faubion, J.D. (ed), Penguin Books, London 2002.

Hobbes, T., *Leviathan*. Cambridge University Press, Cambridge, 1996.

Koselleck, R., *Critique and Crisis: Enlightenment and the Pathogenesis of Modern Society*. The MIT Press, Cambridge, Mass., 1988.

Mansfield, N., 'Fighting for Peace: From the Social War to Armed Democracy'. *The Metamorphosis of War*. Plaw, A. (ed), Rodopi, Amsterdam and New York, NY, 2012.

Mouffe, C., *The Return of the Political*. Verso, London, 1993.

—, *On the Political*. Routledge, London, 2005.

Schmitt, C. *The Concept of the Political*. Trans. Schwab, G., University of Chicago Press, Chicago, 1996.

—, *Political Theology: Four Chapters on the Concept of Sovereignty*. Trans. Strong, T.B., University of Chicago Press, Chicago, 2005.

—, *The Crisis of Parliamentary Democracy*. Trans. Kennedy, E., The MIT Press, Cambridge, Mass., 1988.

Sieyes, E., 'What is the Third Estate?'. *Political Writings*. Sonenscher, M. (ed), Hackett, Indianapolis, 2003.

Von Clausewitz, C., *On War*. Trans. Graham, J.J., Penguin Books, London, 1982.

Jason Edwards is Lecturer in Politics, School of Politics and Sociology, Birkbeck, University of London. He is author of *The Radical Attitude and Modern Political Theory* (Palgrave MacMillan, London, 2007).

Why There is No Such Thing as Political Terrorism

Bob Brecher

Abstract
Naming as 'terrorism' the actions of the various political organisations attempting to effect change through violence aimed at terrorising their targets, both directly and indirectly, and thus at causing them to bring pressure to bear on their political masters to change course, is at once to misdescribe the pursuit of specific political ends as the pursuit of terror for its own sake and to conceal the terroristic nature of actions *not* thus described. 'Terrorism' names no genuine ideology: as the typical characterisation of terrorists as agents of destruction whose only objective is destruction itself implicitly recognises, terror*ism* would be a matter of creating, inducing or spreading terror *as an end in itself*, rather than as a means to some other end. But that is not what any political terrorist group is doing. To talk of 'terrorism' misconstrues a practical and/or rhetorical tactic as a substantive position; effectively mislabels opponents of neo-liberalism; and enables western governments to ensure that the public literally does not see that their own actions in Afghanistan and Iraq are themselves acts of terror on a massive scale.

Key Words: Terror, terrorism, terrorists.

*****

1. Introduction

The US military doctrine governing its war on Iraq tells us that 'Shock and Awe are actions that create fears, dangers, and destruction that are incomprehensible to the people at large, specific elements/sectors of the threat society, or the leadership.'[1] That is as good a description of terrorist acts as any, not least as it has the advantage of not insisting by fiat that such actions have to be directed deliberately at innocent people in order to count as such: for terrorist acts are designed to provoke terror, and terror is something to which anyone may be subjected, whether schoolchild, office worker, cleaner, tourist, soldier or politician.[2] They may be terrorised deliberately, as in the London bombings of 7 July 2005; or they may be terrorised unintentionally, that is to say, without being deliberately targeted. Since anyone may be terrorised, to insist that only those actions that deliberately target innocents are to be described as terrorist is, first, simply to attempt to stipulate rather than to justify what is to count as a terrorist act and, second, to run the considerable risk that that will 'contribute to toleration of the killing of *more* innocents.'[3] For on such an account, and given the distribution of power—the power to name, as well as the power to exercise violence—in the world as it is, terrorising non-innocents is not to commit an act of terror; so the category of 'non-innocent' expands to include people who, whatever their views or commitments, remain non-combatants. Consider for instance the Israeli army's treatment of Palestinian civilians in Gaza, as described by Physicians for Human Rights—Israel.[4] It should not surprise us that, in circumstances like these, terrorist actions, whether deliberately targeting non-innocents or affecting them as 'collateral damage'—and however much that description may be resisted—will become more likely to be tolerated, and thus more often committed.

It is from this perspective—that terror is something familiar, that it is ubiquitous both historically and socially—that I shall argue that, apart from politically quite irrelevant individuals or groups whose end really is simply to terrorise people, *there is no terrorism*. Certainly we are living in a time of terror: just ask the people of Palestine, Afghanistan or Iraq. But that is nothing new. For decades, people around the world have been living in a time of terror: mental patients in North America and the UK in the 1950s and beyond, the people of Chile from 11 September 1973 until their liberation from Pinochet, the people of Argentina under the military junta from 1976 to 1983, and countless others since 1945—in Bolivia, Ecuador, Guatemala, Honduras, Thailand, Indonesia, Cambodia, China, Sri Lanka, Burma, Algeria, Tunisia, Kenya, Congo, Rwanda, South Africa, Bosnia, Kosovo and so on and on. It is not to diminish or deny their nature to point out that the scale and scope of the terrorist actions that 'we' have recently been experiencing in Europe and the United States are minuscule by comparison. But it is not just that the arrogant neo-colonial assumptions governing who 'we' are need to be rejected, as though 'our' lives were of a different order from 'their' lives. Even more importantly, the very notion of 'terrorism' as invoked by our politicians and their corporate masters is a lie; and not just a lie, but a lie which in the same moment that it deceives the populace about the so-called terrorist threat also deflects any criticism of our masters' and mistresses' actions as no less acts of terror than those of, for example, al Qaeda.

2. Terrorism and Terrorism as 'An End in Itself'

Let me start by clarifying the distinction between political terrorism—that is to say, the use of terror in the pursuit of political ends (under which head I shall here include religious ends insofar as these impact on secular politics[5])—and terrorism as a so-called end in itself. Now, I am not sure what the idea of an end in itself, or an end pursued for its own sake, actually amounts to in most cases where it is invoked. It seems to me that most activities thus described are in fact undertaken on account of their causing, leading to or instantiating some further state of affairs; and it is not until we reach a very long way down, so to speak, that this sort of chain comes to an end. Consider what is probably a paradigm of an end in itself: the enjoyment of art 'for art's sake.' It is not a question of the monetary value of the painting, or its serving as a status symbol, or anything at all like that, but rather one of its intrinsic aesthetic value; it is simply a wonderful painting, quite regardless of any end that its being so might serve and/or regardless of its being recognised as such. Think for example of Picasso's *Guernica*. But does not the act of appreciating it 'in itself' have some connection with, say, enjoyment? Could a work of art be wonderful, beautiful or whatever 'in itself' in the absence of anyone's enjoying its wonder or beauty? Would it remain 'beautiful in itself' in a world empty of human beings? Well, perhaps so: certainly Plato would have thought so, Beauty being one among the eternal Forms. But only if you are prepared to go all the way down, with Plato, to the Forms. Or think of someone's snowboarding 'for its own sake.' Is there really no pleasure, or set of pleasures, to which that activity gives rise? Is it not exhilaration that is the end here? Perhaps, then, it is that pleasure, or set of pleasures, that is being pursued 'for its own sake;' 'I just love the buzz that snowboarding gives me.' Still, might one not intelligibly ask someone why they're doing something that gives them the buzz they love? After all, ascetics might eschew just that, because they do not wish to be the sort of person who does whatever it may be that they love to do. Nor is their asceticism itself necessarily something pursued 'for its own sake;' rather, they may well have a vision of the sort of person they ought to try to be, which is why it is asceticism that they pursue, rather than art or snowboarding. Somewhere around this point, perhaps, the chain stops: but it is certainly a pretty deep point, this business of being this and not that sort of person. Or consider Roman Catholic priests and sex, at least in some cases if clearly not in all; they remain celibate, not 'for the sake of being celibate,' but because celibacy is thought to serve certain religious ends. Do they in turn pursue those ends 'for their own sake,' or for God's sake? I do not presume to know: but either way, what they do 'for its own sake,' or what they pursue as 'an end in itself'—the point at which there is no further end that an activity serves—is again a long way down the line.

That alone should give us pause about the notion of 'terrorism for its own sake' or 'terrorism as an end in itself.' Even figures such as Timothy McVeigh (the Oklahoma bomber, 1995) or the perpetrators of school or university massacres such as those at Dunblane (1996), Columbine (1999) or Virginia Tech (2007) would seem to have some sort of political, social or related consciousness—grudge, perhaps—that led them to do what they did.[6] Nonetheless, it is possible at least to imagine a person's committing an act of terror with no further end in mind—or at least, in light of the discussion above, with no further *political* end. Think here, perhaps, of figures such as Charles Manson. Furthermore, the distinction between political terrorism and terrorism for its own sake is not empirically fixed or rigid. First, political terrorists whose goals remain year after year as far away as ever from being realised might cease to have any clear political goal but still carry out acts of terror, if not out of habit, then as a gesture expressing their political irrelevance. Alternatively, I suppose, some deranged individuals who enjoy terrorising others—think of some, but not other, members of the SS, or of certain perpetrators of domestic violence—might come to imagine or invent some (quasi-)political framework within which they do what they do. (Groups are a different matter, of course, just because it is likely to be some sort of political, or religious, beliefs that identify a group as a group in this sort of context.) Second, there are cases that do not clearly fall into either category from the outset. It is unclear, I think, whether or not the Aum Shinrikyo sect responsible for the 1995 sarin attacks on the Tokyo underground had any political goal or not. The same is true of the man who ran amok with a machete in a London nursery school a few years ago: just what was the difference, so far as he was concerned, between a political grudge and personal 'issues?' I do not know, and I suspect that no one else does either. But neither of these limitations on our ability to distinguish political terrorism from these sorts of cases present a serious difficulty for the distinction.

Only if it were thought that definition was to the point would such a problem arise. Definition, however, is not at all to the point, since real things and phenomena can only be described, not defined—just because they are liable to change.[7] A circle, for instance, is defined as a plane figure bounded by a single line every point of which is equidistant from a point at the centre of the figure. Change the definition and you change what is defined; it is no longer a circle that you are defining. The wheel on the front of that bicycle over there, however, cannot be defined, but only described. It may rust, buckle, become scratched, be removed from the bicycle and so on; yet, for all these changes, it remains that wheel. The best that may be said is that such and such characteristics, or set of characteristics, are enough for that thing to be properly described as a wheel; anything like that is sufficient to make the thing a wheel. What is necessary for that to be the case, however, can and does change: how buckled may the wheel be before it is no longer a wheel, or how many spokes are necessary for it to remain a wheel? Unlike in the case of circles, we decide what counts, and what continues to count, depending on our particular and contingent purposes—for instance, using the thing as a wheel for the bicycle we are constructing or as an example in a language lesson of what 'wheel' describes in English. The point is, as Jason Edwards rightly insists elsewhere in this volume, that we would tend to see acts which have destruction and unspeakable cruelty etc. as their overriding goal as the action of madmen. And while such acts may be acts of *terror*—they may terrify and/or terrorise both those at whom they are aimed and others—they do not constitute acts of *terrorism*. Indeed, insofar as they really are the acts of people who are mad, they are barely, if at all, *acts*, just because they are not the outcome of rationally viable intentions. What mad people 'do' when they are 'doing' something that leads others to describe them as mad is to *behave* in such and such a way; their 'actions,' being unintelligible, are not actions at all, but merely behaviour. Consider how we talk about kleptomaniacs or, sometimes, violent schizophrenics. To the extent that they really are not responsible for what they 'do,' they're not doing it; rather, something is happening to them. That is why we do not in fact hold them responsible for what they appear to be doing, since they are not really doing anything, any more than the wind is doing something to the branches and leaves of a tree when it moves them.[8] People who employ terror as a weapon in their pursuit of certain ends are just as different from people who are mad as are those people who rationally pursue their ends by quite different means.

3. What Political Terrorism Is and What It Is Not

It is political systems and ideologies that are rightly named conservatism, liberalism, socialism, anarchism, despotism and so on. The suffix 'ism' denotes a 'doctrine, theory or practice having, or claiming to have, a distinctive character,' according to the *Oxford English Dictionary*. Nor is that merely a semantic point, or one peculiar to English.[9] As the typical characterisation of terrorists as 'agents of destruction whose only objective is destruction itself'[10] implicitly recognises, terror*ism* would be a matter of creating, inducing or spreading terror *as an end in itself*, rather than as a means to some other end (again, with the earlier caveats in place about the notion of an end in itself). But that is not at all what any political terrorist group is trying to do, and thus not what any actually existing terrorist group—and certainly those against which the 'war on terror' is said to be fought—is trying to achieve. The beliefs they are trying to further through terrorist acts may be political, quasi-religious or both (bearing in mind in particular that that distinction itself is not one which may be universalised in any simple fashion). It is not their beliefs about the efficacy of terror that they are trying to spread, nor terrorist acts themselves—apart from the matter of provoking state terror, of course, although even that is a matter of tactics, not of pursuing 'destruction itself.' These groups have clear political objectives; and so do the governments at which their terrorist acts are ultimately aimed. 'Terrorism' names no political ideology.

Consider these comments:

> a war for freedom and against tyranny … a war against those who favour death and by those of us who favour life. We are fighting against nihilists, against agents of

> destruction whose only objective is destruction itself, although they disguise this with social crusades.[11]

Despite appearances, this is not a call to arms in today's so-called war on terror. It is an assessment of, and an attack on, the democratic opposition to military dictatorship in Argentina by Admiral Massera, one of the ruling triumvirate of the military junta from 1976-1983. Yet it is exactly this sort of myth that is now being at once created, perpetuated and mobilised by the Bush regime and its international supporters, the myth that 'We are fighting against nihilists, against agents of destruction whose only objective is destruction itself.' Whatever we may think of the political objectives of the Tamil Tigers, al Qaeda or the various Palestinian terrorist organisations, 'destruction itself' is not among them. They are very far from being 'nihilists.' Their political objectives are not those of, for instance and arguably, late nineteenth-century anarchist groupuscules; and of course they are by no means devoid of political vision. Quite the contrary: they are entirely positive (in a neutral sense of the term, and whatever your view of their political programme): the reordering of the Sri Lankan state; the end of American occupation of the heartland(s) of Islam and a viable Palestinian state. Certainly, they use terror as a means to their ends. But that is entirely different from putatively pursuing it as some sort of end in itself, from terror *per se*.

Comparing the idea of terrorism with that of extremism helps to make this clear. Simply to be devoted to extremes for their own sake, whatever they are and wholly independently of any ends such devotion might serve, is unintelligible. What would an *ideology* simply of extremism look like? What would its substantive content be? To elevate what is, at best, a practical and/or rhetorical tactic to a substantive position—to elevate it, if you like, from adjective to noun—is sheer obfuscation. It is effectively to label as an extremist anyone who opposes the neo-liberal vision of the world, as if that were uncontestably the norm, rather than its being a norm that its proponents are seeking to impose on others. Or consider torture. Doubtless there are sadists among the world's torturers who torture for the pleasure of it: but that is entirely different from the use of torture to intimidate or terrorise a population. There is no 'torturism,' a few individual sociopaths perhaps apart: not even the torture at Abu Ghraib was 'torturism,' torture for its own sake, torture in some bizarre quasi-Sadeian sense as describing a vision of the good life.[12] As with torture and extremes, so with terror: all are means of attaining certain ends; they are not '-isms.' They name no practically possible ideology.

But of course to characterise the actions of 11 September 2001, or those of Palestinian militants in the continuing intifada, as 'terrorism' allows those in command, both of language and of the state and its organs, to persuade people that such terrorists are simply 'the evil ones [who] have roused a mighty nation,'[13] people who are not rational political agents but agents of 'evil, the very worst of human nature.'[14] In turn, that is what enables them to conduct 'a war against evil people.'[15] It allows those who command the world to ignore both the terrorists' objectives and, even more importantly, the causes of their having those objectives and their reasons for adopting them; to command that these be sidelined or ignored, and thus to command the meanings of the terms at issue—and of 'terrorism' in particular. It allows them to substitute for the 'Evil Empire' of the later twentieth century a new enemy. To the extent that the democracy in whose name they rule is not merely a formal front for a more controlled version of the Hobbesian 'state of nature,' one may be forgiven for supposing that, as Nick Mansfield argues in his contribution to this volume, democracy is not at all the innocent mere mechanism it is generally taken to be. Perhaps democracy requires an 'other' in order to delimit its possible scope, so that where no suitable 'other'—such as the Soviet Union—presents itself one has to be invented. Or to put it another way, when democracies run out of their supply of enemies—often of course former friends later turned enemies, such as Sadaam Hussein or the Taliban—they simply invent new ones. Jason Edward's reminder elsewhere in this volume is indeed a timely one: 'Today's enemies may be tomorrow's allies, and our allies in today's victory may be our enemies tomorrow.'[16]

Furthermore, in conditions of the disobedience, whether real or actual, of a great many of the benighted people across the globe who do not enjoy the privileges and pleasures of democracy, it will not just be the case that war serves as democracy's double, but that, increasingly, it will be terror that does so.[17] As the real alternative to democracy—which is how terror is presented—it serves to discipline the citizens of nation-states on two essential fronts at the same time: the external threat allegedly presented, and represented, by the so-called terrorism of groups such as al Qaeda; and the real internal threat of terror, in the shape of 'anti-terrorism' laws, offered by their governments to the citizens of democratic polities should they be tempted to disobey or even unduly to question the neo-liberal realities of actually existing democracy. Democracy as we know it certainly seems to require 'others,' or enemies, no less than any other system of government; war, far from its antithesis, is essential for it to function as it does; and it is terror and the threat of terror that protects that functioning. Whether this is the case only of actually existing democracies, or of any possible democracy, the question Mansfield begins to explore, is a crucial question of our times. No less crucial, albeit in an entirely parochial context, is the fact that it is no exaggeration to say that terror is quite clearly democracy's double in today's United Kingdom. The United Nations' Committee on Human Rights' coruscating report on the UK makes this abundantly clear. Among many other things, it unequivocally condemns the fact that terrorism legislation, and in particular the 2006 Terrorism Act, has undermined human rights in the UK: to take just one example

> a person can commit the offence 'encouragement of terrorism' even when he or she did not intend members of the public to be directly or indirectly engaged by his or her statement to commit acts of terrorism, but where his or her statement was understood by some members of the public as encouragement to commit such acts.[18]

That democracy's double serves in fact to subvert it may or may not point to a deep internal contradiction in the very idea of democracy. However that may be, it is clear that, with its anti-terrorist legislation, the UK government is seeking quite deliberately to terrorise the people it purports to represent, and that it does so in the name of the defence of just those democratic practices and values it is in fact destroying.

In seeking to persuade us that bin Laden was some sort of barbarian madman dedicated to terror for its own sake, that he was someone whose 'attacks of September 11 drew a bright line of demarcation between the civil and the savage,'[19] the leader of a bunch of crazed fanatics, 'the depth of [whose] hatred is equalled by the madness of the destruction they design;'[20] our leaders have two inextricably linked objectives. First, and most obviously, such a characterisation deflects attention from the fact that what al Qaeda condemns the United States for is simple: its policies in the Middle East. Second, and even more perniciously, this presentation and characterisation of terrorism as somehow a movement and/or an ideology enables the American administration and their allies to ensure that the public overlooks, that it literally does not see, what is clearly there to be seen: namely that its own actions in Afghanistan and Iraq are themselves acts of terror on a massive scale, and the latest in a long line of terrorist acts marking the pursuit of American foreign policy since 1945. If it is the case that al Qaeda and its supporters 'hate all civilization and culture and progress' and 'have divorced themselves from the values that define civilization itself,'[21] then what is one to say of the USA, the UK and their allies? If it is indeed the case that committing acts of terror places the people responsible—'freedom's enemies, murderers of innocents in the name of a barbarous cause'—on 'one side' of 'the bright line of demarcation,' while the 'friends of freedom' remain '[o]n the other side,'[22] then how are we to understand and evaluate the bombing of Iraq, the sacking of Fallujah, the torture practised at Guantanamo Bay and in secret CIA torture chambers around the world and countless other acts of terror? Following this 'logic of "terrorism,"' and on the basis of the principle, 'like effect, like cause,' we would have to conclude that, the evidence being the same—the murder 'of innocents in the name of a barbarous cause' (namely that of the Chicago School economists, the corporations and the corporate-statist politicians who sustain the neo-liberal push for hegemony[23])—then what is evidenced must also be the same. The 'friends of freedom,' on the evidence of their actions, are in fact 'freedom's enemies.'

So they are, even though their own use of terror to achieve their political ends no more betokens their dedication to terrorism than it does in the case of al Qaeda. What matters is the preparedness to terrorise people in the pursuit of one's ends; and that is exactly what the rhetoric of terror*ism* helps to conceal.

4. Conclusion

As Richard Jackson insists, 'the act of naming things is always a highly charged process that can have serious political and social consequences;'[24] and that is because 'For us to live and die properly, things have to be named properly.'[25] Naming as 'terrorism' the actions of the various political organisations attempting to effect change through violence aimed at terrorising their targets, both directly and indirectly, and thus at causing them to bring pressure to bear on (or to withdraw pressure from) their political masters to change course—whether the Americans in the Middle East or the Israelis in Palestine—is at once to misdescribe the pursuit of specific political ends as the pursuit of terror for its own sake and to conceal the terroristic nature of actions *not* thus described. It at once enables those doing the naming to enforce views of the world that are simply erroneous, but belief in which (certainly others' beliefs, but perhaps their own too) is a necessary condition of the pursuit of their own policies. It allowed Blair to make statements such as the following:

> If it is the plight of Palestinians that drives them, why, every time it looks as if Israel and Palestine are making progress, does the same ideology perpetrate an outrage that turns hope back into despair?
>
> If it is Afghanistan that motivates them, why blow up innocent Afghans on their way to their first ever election? If it is Iraq that motivates them, why is the same ideology killing Iraqis by terror in defiance of an elected Iraqi government?[26]

My concern here is not that what Blair said is, in my of course contestable view, in various respects false, important though that is. Rather I want to emphasise that it is the concept of 'terrorism' underlying his questions that enables those questions to do their work; and that their work is to both regard and treat as merely rhetorical the questions he asks, questions to which the implied answer is all too clearly something like 'Because they're crazy.' Even without any factual disagreement, we are being inveigled at once into supposing three things. First, that Blair's question about innocent Afghans can have no rational answer (however unpalatable—but that is another matter). Second, he is trying to persuade us that, since the perpetrators of the attacks concerned are irrational, barbaric, uncivilised or crazed, there can be no question of comparing their acts of terror with those of the UK government, for the latter is 'of course' not any of those things. Third, we are being asked to suppose precisely what is not the case, namely that terrorists are necessarily irrational, that their actions instantiate insanity. It is this sort of silencing and lying that the language of 'terrorism' helps at once to encourage and secure.

Terrorism, the pursuit of terror for its own sake, makes no sense; it is irrational, evil, incomprehensible. 'Their' terrorist actions are therefore entirely beyond understanding, perhaps even uniquely evil; and so we come to understand ourselves to be living at a unique time, a time of terror, a time in which any anti-terrorist measures—however corrosive of the very values they purportedly defend—are justified as a matter of defending our own essential interests. Furthermore, since our actions, even those resulting in inevitable but regrettable 'collateral damage,' being actions that do make sense, actions that are not irrational, evil or incomprehensible, simply *cannot* constitute terroristic acts.[27] 'We,' then, remain innocent; and what 'we' do in the world is a matter of regrettable necessity, whether at Guantanamo Bay, in Fallujah or in America's surrogate torture chambers around the world. The language of 'terrorism' helps prevent us from realising precisely what we need to grasp if we are to adopt policies that will lead to fewer, rather than more, acts of terror, namely that '[T]he precondition for thinking politically on a global scale is to see the *unity* of the unnecessary suffering taking place.'[28]

John Berger is right to insist that '[W]e have to reject the new tyranny's discourse. Its terms are crap.'[29]

Notes

[1] B. Edney, 'Appendix A: Thoughts on Rapid Dominance,' *Shock and Awe: Achieving Rapid Dominance*, H.K. Ullman and J.P. Wade (eds), NDU Press Book, Washington DC, 1996, p. 110.
[2] In this regard see in particular T. Honderich, *After the Terror*, *2nd ed.*, Edinburgh University Press, Edinburgh, 2003 and *Terrorism for Humanity*, Pluto, London, 2005.

[3] T. Honderich, *Humanity, Terrorism, Terrorist War*, Continuum, London, 2006, p. 94.
[4] Physicians for Human Rights – Israel, *Holding Health to Ransom: GSS Interrogation and Extortion of Palestinian Patients at Erez Crossing* (August 2008), Accessed 17 August 2008, available at: <www.phr.org.il/phr/article.asp?articleid=603&catid=55&pcat=-1&lang=ENG>.
[5] It is in their impact on our lives here on Earth (and whether or not such lives are all the lives there are or can be) that such religious, semi-religious or quasi-religious beliefs and goals of, for instance, born-again Christians expecting the Rapture or Muslims seeking to institute a Middle Eastern or world-wide caliphate are significant, at least so far as those are concerned who do not share in such, or indeed in very different, religious visions or whose interest in them is not in some sense theological.
[6] With thanks to Jason Edwards for pointing this out.
[7] Cf. torture: B. Brecher, *Torture and the Ticking Bomb*, Blackwell, Oxford, 2007, pp. 3-6.
[8] I realise of course that this is anathema to Freudians; but that is not something to be pursued here. The substantive point, that people who are mad and whose irrational behaviour (me) or (unconscious) actions (Freud) terrorise people are not on that account to be described as engaging in terrorism. Thanks to Angela Fenwick for pushing the point.
[9] French, '-isme', German and Czech, both '-ismus', for example, are the same.
[10] See following note.
[11] Admiral Massera, quoted by M. Feitlowitz, *A Lexicon of Terror: Argentina and the Legacies of Torture*, Oxford University Press, New York, 1998, p. ix.
[12] See for example P. Gourevitch and E. Morris, *Standard Operating Procedure*, Picador, Basingstoke and Oxford, 2008.
[13] G.W. Bush, Press Conference, The East Room, Washington DC, 11 October 2001; with thanks to R. Jackson, *Writing the War on Terrorism*, Manchester University Press, Manchester, 2005, p. 66.
[14] G.W. Bush, 'Statement by the President in His Address to the Nation,' 11 September 2001; with thanks to Jackson, op. cit., p. 67.
[15] C. Rice (National Security Advisor), Interview by Al Jazeera Television, 15 October 2001; with thanks to Jackson, Ibid.
[16] J. Edwards, 'Foucault and the Continuation of War,' *The Metamorphosis of War*, A. Plaw (ed), Rodopi, Amsterdam and New York, NY, 2012, p. 31.
[17] N. Mansfield, 'Fighting for Peace: From Social War to Armed Democracy,' *The Metamorphosis of War*, A. Plaw (ed), Rodopi, Amsterdam and New York, NY, 2012, pp. 13-14.

[18] United Nations International Convention on Civil and Political Rights, CCPR/C/GBR/10/6, 30 July 2008, Sec.26, Accessed 16 August 2008, Available at: <http://www2.ohchr.org/english/bodies/hrc/hrcs93>.
[19] J. Ashcroft (Attorney General), 'Testimony to House Committee on the Judiciary,' 24 September 2001; with thanks to Jackson, op. cit., p. 62.
[20] G.W. Bush, 'The President's State of the Union Address,' 29 January 2002; with thanks to Jackson, Ibid.
[21] G.W. Bush, 'Remarks by the President to the CEO Summit,' Pudong Shangri-La Hotel, Shanghai, People's Republic of China, 20 October 2001; with thanks to Jackson, Ibid., p. 49.
[22] J. Aschcroft, op. cit.; with thanks to Jackson, Ibid.
[23] See N. Klein, *The Shock Doctrine*, Allen Lane, London, 2007, for an excellent discussion of the ways in which violence is integral to advancing the neo-liberal cause.
[24] Jackson, op. cit., p. 23. Cf. D. Spender, *Man-Made Language*, Routledge & Kegan Paul, London, 1980.
[25] J. Berger, *Hold Everything Dear*, Verso, London, 2007, p. 42.
[26] Full text of 'Blair Speech on Terror,' BBC News 24, Accessed 16 July 2005, available at: <http://www.news.bbc.co.uk/1/hi/uk/4689363.stm>.
[27] There is of course a variously unfortunate parallel here between the claim that 11 September 2001 'changed everything' and the equally misleading claim that the Holocaust was unique. See for example *Is the Holocaust Unique?*, A.S. Rosenbaum (ed), Westview Press, Boulder, 1996 and my 'Understanding the Holocaust: The Uniqueness Debate', *Radical Philosophy* Vol. 96, 1999, pp. 17-28.
[28] Berger, op. cit., p. 38.
[29] Ibid., p. 41.

Bibliography

Berger, J., *Hold Everything Dear*. Verso, London, 2007.

'Blair Speech on Terror'. BBC News 24.16 July 2005, viewed 3 January 2006, <http://www.news.bbc.co.uk/1/hi/uk/4689363.stm>.

Brecher, B., *Torture and the Ticking Bomb*. Blackwell, Oxford, 2007.

—, 'Understanding the Holocaust: The Uniqueness Debate'. *Radical Philosophy*. Vol. 96, 1999, pp. 17-28.

Edney, B., 'Appendix A: Thoughts on Rapid Dominance'. *Shock and Awe: Achieving Rapid Dominance*. Ullman, H.K. and Wade, J.P. (eds), NDU Press Book, Washington DC, 1996.

Edwards, J., 'Foucault and the Continuation of War'. *The Metamorphosis of War*. Plaw, A. (ed), Rodopi, Amsterdam and New York, NY, 2012.

Feitlowitz, M., *A Lexicon of Terror: Argentina and the Legacies of Torture*. Oxford University Press, New York, 1998.

Gourevitch, P. & Morris, E., *Standard Operating Procedure*. Picador, Basingstoke and Oxford, 2008.

Honderich, T., *After the Terror, 2nd ed.* Edinburgh University Press, Edinburgh, 2003.

—, *Terrorism for Humanity*. Pluto, London, 2005.

—, *Humanity, Terrorism, Terrorist War*. Continuum, London, 2006.

Jackson, R., *Writing the War on Terrorism*. Manchester University Press, Manchester, 2005.

Klein, N., *The Shock Doctrine*. Allen Lane, London, 2007.

Mansfield, N., 'Fighting for Peace: From the Social War to Armed Democracy'. *The Metamorphosis of War*. Plaw, A. (ed), Rodopi, Amsterdam and New York, NY, 2012.

Rosenbaum, A.S. (ed), *Is the Holocaust Unique?* Westview Press, Boulder, 1996.

Spender, D., *Man-Made Language*. Routledge & Kegan Paul, London, 1980.

United Nations International Covenant on Civil and Political Rights: Human Rights Committee, 93rd Session, Geneva 5 – 25 July 2008, 'Consideration of Reports Submitted by States Parties Under Article 40 of the Covenant: Concluding Observations of the Human Rights Committee, United Kingdom of Great Britain and Northern Ireland,' 2008.

Bob Brecher is Professor of Moral Philosophy at the University of Brighton, and Director of its Centre for Applied Philosophy, Politics & Ethics. He has published over sixty articles in moral theory, applied ethics and politics, healthcare and medical ethics, sexual politics, terrorism and the politics of higher education, as well as several authored and edited books. His latest book, *Torture and the Ticking Bomb* (Blackwell, 2007) is the first book-length rebuttal of calls to legalise interrogational torture. Currently he is working on a theory of morality as practical reason, building on his earlier *Getting What You Want? A Critique of Liberal Morality* (Routledge, 1997).

Part II:

Confronting the New Wars: Law, Security and Diplomacy

The Legality of Targeted Killing as an Instrument of War: The Case of the US Targeting of Qaed Salim Sinan al-Harethi

Avery Plaw

Abstract:
This chapter examines the legality of US targeted killing operations under international law by focusing on one well-established case, the targeting of Qaed Salim Sinan al-Harethi in Yemen on 3 November 2002. It assesses the operation on the basis of three potentially applicable legal frameworks—criminal law enforcement, self-defence short of war and the humanitarian law of armed conflict. It concludes that while there may be a plausible case for the legality of the al-Harethi operation, this case relies on elements that have not necessarily been present in other US targeting operations. The al-Harethi case thus helps to illuminate some of the conditions that would be required to carry out legitimate targeting operations.

Key Words: Targeted Killing, terrorism, self-defence, humanitarian law, criminal law, military operations.

*****

1. The Yemen Operation

On 3 November 2002 a CIA-controlled Predator unmanned aerial vehicle appeared over a car speeding along an isolated highway 100 miles east of Sanaa, the capital of Yemen. The Predator launched a laser-guided hellfire missile which struck the target and exploded, leaving only charred remains. American and Yemeni officials claimed that the cars' occupants had been six dangerous terrorists including Qaed Salim Sinan al-Harethi, an al-Qaeda leader the CIA believed to be responsible for the bombing of the US destroyer Cole in 2000. Based on the carbonized remains, forensics specialists were unable to confirm the identity of the victims.[1]

Despite the difficulties with identification, Paul D. Wolfowitz, the then US Deputy Secretary of Defence, praised the venture as 'a very successful tactical operation,' and implied that such operations would be a continuing feature of America's new war on terrorism.[2] In fact, American officials now acknowledge having carried out at least 19 successful targeting operations since 11 September 2001.[3]

International reactions to the operation and to others like it have differed sharply. On one hand, the United Nations (UN), the European Union (EU) and many states, non-governmental organizations (NGOs) and legal experts denounced the American operation as an extrajudicial killing or assassination, both banned under international law. For example, the UN Special Rapporteur on Extrajudicial, Summary or Arbitrary Executions condemned the operation in his annual report as 'a clear case of extra-judicial killing.'[4] Swedish Foreign Minister Anna Lindh described the operation as 'a summary execution that violates human rights.'[5] Amnesty International has condemned it as an 'extra-judicial execution.'[6] Several influential legal scholars have argued that the action also constituted 'assassination.'[7] For example, Helen Duffy, the Legal Director of the INTERIGHTS, has characterized the 'Yemen attack... [as] the clearest example' of the US government's new 'policy of assassinations.'[8]

On the other hand, the US and other states and legal experts have argued that the al-Harethi targeting was a legitimate exercise of self-defence under international law.[9] Indeed, some of the most influential scholars of humanitarian law have defended these types of operations.[10] Moreover, in December 2006 the Israeli Supreme Court sitting as the High Court of Justice upheld the legality of terrorist targeting in certain circumstances.[11] Even the Director of Human Rights Watch, Kenneth Roth, has defended the targeting of al-Harethi, remarking that 'treating him as an enemy combatant and resorting to lethal force was appropriate.'[12] Moreover, the reluctance of much of the international community to condemn the American action in Yemen has been widely noted.[13] Finally, a number of scholarly articles have recently been published arguing that in some cases targeting operations may be defensible not only legally but also morally.[14]

This chapter examines the legality of targeted killing under international law by focusing on this first clear and unambiguous US operation. It evaluates the reported facts in terms of three potentially applicable bodies of law—criminal law, international law relating to self-defence short of war and international humanitarian law. It concludes that there may be a plausible case for the legality of the al-Harethi operation. However, the case for the legality of the US action relies on elements that may not be applicable to many other targeting operations. The al-Harethi case thus helps to clarify the legal requirements and limits of targeted killing.

Several legally pertinent aspects of the Yemen operation warrant particular attention. In listing these alleged facts, however, it should be emphasized that they are based on coverage in the popular press, NGO reports and academic journals. They have not been subject to the degree of scrutiny that would apply to evidence at a trial. Nonetheless, it appears that:

- the US acted with the permission of Yemen.[15]
- There was little doubt that al-Harethi was a senior al-Qaeda leader. The US also claims that he had been a central figure

in the attack on the USS Cole. The US has not made the evidence public, but the claim has been widely accepted.

- The other passengers in the SUV were either al-Qaeda members or known members of a terrorist organization closely affiliated with it, the Aden-Abyan Islamic Army. One passenger was also a US citizen—Kamal Derwish, an Arab-American who grew up near Buffalo and, according to the FBI, had recruited Americans for terrorist training at al-Qaeda camps.[16]
- No bystanders were harmed, nor, given the remote location of the strike, does it appear that there was a significant danger of such injuries.
- The Yemeni government reported several failed attempts to arrest al-Harethi.[17] Indeed, a failed December 2001 attempt to arrest al-Harethi is reported to have resulted in the deaths of twenty people (including at least twelve Yemeni soldiers at the hands heavily armed tribesman allied with al-Harethi).[18]
- The wreckage of the car revealed remnants of communication equipment and traces of explosives. It has also been reported that authorities 'suspected that the passengers [in al-Harethi's vehicle] were going to a target.'[19]

There are at least three legal paradigms that could plausibly be applied to the targeted killing of al-Harethi: (i.) criminal law, (ii.) self-defence short of war and (iii.) the humanitarian law of armed conflict. Each legal paradigm has its own implications for the legality of the operation. They will be considered in order below.

2. The Criminal Law Enforcement Paradigm

In November 2002, Yemeni officials reportedly affirmed that they were working 'in co-ordination' with the US.[20] It appears then that the Yemeni government at least consented to the US operation. The operation was therefore not an act of international aggression. Indeed, one way of interpreting it would be as a use of force at the invitation of, and in support of, the Yemeni government.

If the Yemeni government invited US action, then there is a well-established legal framework in relation to which the al-Harethi targeting might be justified. There is little doubt that it is legal for a country that is having difficulty in enforcing criminal law at home to call upon the aid of another. This point was explicitly recognized by the International Court of Justice in the case of *Nicaragua v. United States of America.*[21]

However, in a case of invited intervention, the foreign country's right to use force would extend no further than the local government's authority. In other words, the country inviting action could not authorize the intervening power to use force in ways that it was itself not permitted.

At the time of the al-Harethi targeting, there was no situation of armed conflict in Yemen that would warrant the Yemini government in resorting to lethal military operations within its own territory. The government of Yemen therefore had a right to pursue terrorists, but only within the framework of criminal law. The Yemeni government then only had the authority to use lethal force against al-Harethi if it was 'strictly unavoidable to protect themselves or other persons from imminent threat of death or serious injury.'[22] For example, the Yemeni government could legally use lethal force against al-Harethi if he violently resisted arrest and posed an immediate and substantial threat to bystanders or law enforcement agents.

On the 'invited intervention' interpretation then, US operations in Yemen would have similarly been restricted to the framework of law enforcement. In that framework, the US only had the authority to use lethal force against al-Harethi if he posed an imminent danger of serious harm to others and there was no other means to neutralize the threat.

It may be argued in defence of the US operation that al-Harethi did pose a substantial threat to others. In fact, al-Harethi had resisted an attempt to arrest him in December 2001, resulting in the death of around 20 persons. Although the incident was almost a year old at the time of his targeting, authorities appear to have been well within their rights to treat al-Harethi as a dangerous fugitive, and to be prepared to use lethal force if necessary.

But was the use of lethal force necessary at the time that it actually was used? More specifically, was the threat that he posed of sufficient urgency to justify resorting to lethal force? After all, the car carrying al-Harethi was driving in an isolated stretch of desert at the time of his targeting, so it is difficult to see who was urgently threatened.

Still, defenders of the US operation may point out that even if al-Harethi did not pose an instantaneous danger at the moment that he was targeted, there appears to be some evidence that he may nonetheless have posed an imminent danger to others in the near future. Although official records are not available to the public, it has been reported that traces of explosives were found in the wreckage of the SUV. If this is so, then there may be reason to believe that al-Harethi was in an advanced stage of planning or even carrying out a terrorist attack. A case may also be made that he was attempting to flee Yemeni jurisdiction. Either or both of these circumstances could be offered as a rationale for the use of force to stop him.

Nevertheless, neither of these explanations are entirely convincing for two reasons: (1.) al-Harethi was not alone in the vehicle, and even if it could be assumed that he would violently resist arrest, the same cannot necessarily be assumed of the other passengers; (2.) moreover, al-Harethi and his companions were not offered any chance to surrender—a possibility that cannot be discounted when menaced with hellfire missiles. In the context of a law enforcement operation, and in the absence to an immediate and direct threat to others, it may be forcefully argued that suspects, and even fugitives, must be given a chance to surrender. If this argument is compelling, then it does not appear that the deliberate and intentional killing of al-Harethi would comply with applicable criminal law.

3. The Paradigm of Self-Defence Short of War

There is, however, a second legal framework that could potentially be applied to the targeting of al-Harethi—that is, (ii.) *self-defence short of war*. The September 11th attack on the United States was widely recognized by states and international institutions (including the Security Council in resolutions 1368 and 1373) as constituting an armed attack triggering a right of self-defence under article 51 of the UN Charter. The right of self-defence permits the limited use of armed force against perpetrators of armed attacks.[23]

However, depending on the gravity of the attack, the use of self-defence may be restrained by a number of preconditions on the use of force. Interpreting the gravity of an armed attack is a grey area in the law and is subject to legitimate dispute. Assuming, however, that the attack is not seen as meeting the level of gravity required to instigate a situation of general armed conflict, the US would be limited to acts of self-defence short of war against al-Qaeda.

Acts of self-defence short of war are legitimate only if they meet certain preconditions. These conditions are not specified in the UN Charter or international conventions but are largely a matter of established practice. There is, correspondingly, disagreement over exactly what preconditions apply. Two conditions, however, have been recognized as constituting customary law, specifically, necessity and proportionality.[24] These terms are not defined in black letter law, but they are interpreted by the leading contemporary commentator on the law of self-defence, Yoram Dinstein, as follows—here with particular reference to terrorist groups having carried out armed attacks:

> a. necessity (involves two conditions): first, 'a repetition of the [terrorist] attack must be expected, so that the extra-territorial law enforcement can qualify as defensive and not purely punitive;'

> second, '[t]he absence of alternative means for putting an end to the operations of the armed bands or terrorists.'[25]
>
> b. proportionality: the state's 'operations are to be directed exclusively against the armed band or terrorists.'[26]

These requirements (of necessity and proportionality) reflect the basic conditions on the limited use of armed force in self-defence. Indeed, as Christine Gray of Cambridge University notes in her *International Law and the Use of Force* (2004), in general state practice these 'are often the *only* factors relied upon in deciding the legality of particular actions.'[27]

Two other preconditions are sometimes cited, although usually one or the other rather than together. These are immediacy and imminence. Dinstein, for example, insists on the condition of immediacy and defines it as follows: 'there must not be an undue time lag between the armed attack and the exercise of self-defence.'[28] He notes, however, that this 'condition ought not to be construed too strictly,' especially in cases of terrorist attack. States must be permitted the time to gather intelligence, pin the blame, find terrorists and deal with them proportionately.

By contrast, Michael N. Schmitt of the George C. Marshall European Centre for Security Studies, insists on a standard of imminence rather than immediacy. Under an imminence standard, he argues, the right to use force in self-defence arises 'at the point at which the threat can last be thwarted effectively.'[29] He also notes, however, that responding to terrorism constitutes a special case. Given the criminal and clandestine character of terrorists groups, the state will 'seldom possess the war plans of their enemy.' In such cases, therefore, 'circumstantial evidence of intent may be sufficient… to justify action.'[30]

Although once again the lack of details and public justification from the US government make it difficult to gauge, it seems plausible that a case can be made that the targeting of al-Harethi met the established conditions for the use of armed force in self-defence short of war, and probably at least one of the additional contested conditions as well—although an element of uncertainty arises in relation to imminence.

The first question that needs to be answered is whether continued attacks could reasonably be expected. There is a strong case that the answer is 'yes,' especially if al-Harethi, as reported, remained an active al-Qaeda operational leader, if al-Qaeda was known to be planning further attacks and if the evidence that the car was carrying explosives (possibly on the way to an attack) is accurate.

The next question (the second concerned with necessity) is whether there were 'alternative means for putting an end to the operations of the armed bands or terrorists.' Following the failure of repeated efforts to arrest al-Harethi with 20 casualties, and despite the deployment of extensive force by the Yemeni state, it was reasonable to conclude that arrest was not a plausible option—except perhaps under highly favourable circumstances. In the absence of such circumstances, it was reasonable, at least in the context of self-defence, to resort to the use of force.

The question raised by the precondition of proportionality is whether a military operation is 'directed exclusively against the armed band or terrorists.' In relation to this criterion, the killing of al-Harethi himself does not raise difficulties, nor does that of the other al-Qaeda member traveling with him. A difficulty could be raised, however, over the Aden-Abyan Islamic Army members. Interestingly, legal critics of the American action have not aggressively pursued this argument. This perhaps reflects recognition that the Aden-Abyan Islamic Army is widely believed to be a close operational affiliate of al-Qaeda and a participant in a range of operations including the attack on the USS Cole and the planned attack on the USS the Sullivans. Moreover, the Aden-Abyan Islamic Army is independently responsible for serious acts of terror including the 1998 kidnapping of 16 Westerners (resulting in the death of four of them).[31] Nonetheless, even if the Aden-Abyan Islamic Army members do not raise a decisive objection in this case, they nonetheless point to a troubling feature of combating terrorists with military force—i.e., it is sometimes difficult to define who should and who should not be treated as a combatant.[32]

If the arguments outlined above are accepted, then it appears that the targeting of al-Harethi can meet the preconditions for the use of force in self-defence clearly established in customary international law. However, the criteria of immediacy and imminence should also be considered because they also play an important role in the discourse over such uses of force.

There seems to be a good case that the al-Harethi targeting could meet a standard of immediacy. Here the key question is whether the 'time lag between the armed attack and the exercise of self-defence' was reasonable. The operation took place a little over a year after the 9/11 attack, and two years after the Cole attack. It is quite plausible that the intervening time was required to trace responsibility for the attacks, to locate the apparent perpetrators, for alternative means to be attempted to apprehend the suspect and, having failed, to then gain permission from the local government for a military operation and to find an opportunity to conduct it in a way that would not endanger civilians (i.e., proportionately).

The issue of imminence is less clear, although again the circumstantial evidence is suggestive. The question raised is whether there was reason to believe that the targeting attack occurred at the last point that a terrorist attack could be 'effectively thwarted.' Here the facts that have been reported do not permit a clear answer—although the fact that al-Harethi was traveling with a group of closely affiliated terrorists and reportedly carrying explosives is suggestive.

In sum then, the targeting appears on balance to meet the requirement for an act of armed self-defence short of war. It meets the core customary requirements, in addition to one contestable additional requirement (i.e., immediacy). Whether it meets the remaining additional criteria (i.e., imminence) is unclear based on the information currently available.

4. The General Armed Conflict Paradigm

The final legal paradigm that could be applied to the targeting of al-Harethi is that of (iii.) *general armed conflict.* The US government claims to be conducting a global war on terror, and specifically identifies al-Qaeda as its principal adversary. Under conditions of general warfare, 'all combatants can be lawfully targeted.'[33]

If the general framework of war is accepted then there seems little doubt that the targeting of al-Harethi was legal. As an operational leader of al-Qaeda, he would certainly qualify as a combatant in a war with al-Qaeda. The only real issue raised is with the members of the Aden-Abyan Islamic Army, and in-so-far as they are closely affiliated with al-Qaeda and have participated in attacks against the US, they probably also qualify as combatants in the context of an armed conflict.

5. Legal Assessment of the Yemen Operation

Was the targeting of al-Harethi then legal? It appears to be legitimate on two of three applicable legal paradigms. This creates a strong foundation for arguing that it was legal—albeit not simply because two is more than one. There is much dispute today between those who think that the struggle against terrorism should be conducted primarily within the framework of law-enforcement, and those who argue that it is best approached as a general war. There is, however, significantly less dispute about the legality of specific operations conducted under the second paradigm (self-defence short of war), provided that they meet the relevant preconditions (and do not otherwise violate humanitarian law). In-so-far as the operation meets the legal requirements of the paradigm of self-defence short of war, a strong case can be made that it was legal. It is likely for this reason that some human rights groups, such as Human Rights Watch, have declined to condemn the operation.[34] Indeed, Kenneth Roth, the Director of Human Rights Watch, suggested that this might be a justifiable instance of targeting.[35]

6. Implications for the Legal Regulation of Terrorist Targeting

However, even if one accepts (provisionally, based on extrapolation from available evidence) that there may have been a strong legal case for the al-Harethi operation; there are reasons to be reserved about extending this finding to other cases of targeting. First, a number of apparent features marked al-Harethi as an especially plausible target: he was an active al-Qaeda leader who was purportedly personally involved in lethal attacks against Americans, he had successfully resisted attempts at arrest by local authorities causing the deaths of soldiers and civilians, and he was killed while (apparently) carrying explosives under suspicious circumstances. Each of these facts contributed to meeting the applicable standard of necessity. It is also important that the operation could be undertaken without posing a significant threat of harming civilians. This circumstance contributed significantly to meeting the applicable standard of proportionality. In addition, the apparent fact that the Yemeni authorities approved the operation was also relevant in terms of avoiding any potential violations of national sovereignty. Yet it is far from clear that these or parallel factors have been present in all US targeting operations.

Indeed, it is not clear what factors the US government itself sees as regulating the legitimacy of such operations. In conducting targeting operations, the US government is operating in a grey area of international law (involving, for example, the difficult issue of when terrorists can properly be regarded as combatants). In the absence of a clear and specific framework, outlining exactly where, when and against whom targeting can be employed, critics have raised worries that such operations may become more of a threat to innocent civilians than the terrorist attacks they are intended to prevent. The UN Rapporteur on Extrajudicial, Summary or Arbitrary executions, for example, has argued that targeting operations could potentially be used against 'any enemies of the state, social misfits, political opponents, or others.'[36] In this context, imprecision carries just the same perils in regards of terrorist targeting that Benjamin Rampp demonstrates it involves for the idea of human security in another chapter in this book—that is, that what was originally intended to protect human beings becomes itself a greater threat to them.

The foregoing analysis does not suggest that targeting operations are necessarily illegitimate. There appears to be a strong case, for example, that the al-Harethi operation was lawful. But it does suggest the need to clarify the applicable rules. The US government should formally acknowledge that it is its policy to conduct such operations under certain circumstances, outline what it regards as the necessary conditions to warrant such operations and be prepared to defend its policy in international legal and political forums.

Second, the preceding discussion of the al-Harethi case points to an urgent need for more extensive independent oversight of targeting operations.

Even if it is (provisionally) accepted that there may have been a plausible legal justification for a targeting operation in al-Harethi's case, critics effectively point out what it would mean to make a mistake about this question: as Anna Lindh put it, such a mistake would render a targeting operation 'a summary execution that violates human rights.' Even in the comparatively straight-forward al-Harethi case, some question arose over whether members of the Aden-Abyan Islamic Army were properly treated as legitimate targets (in the absence of alternate means of preventing further attacks). I suggested that in light of their purported affiliation with al-Qaeda and their alleged involvement in attacks on Americans, they probably could be treated as combatants in this case. But what about a case where a group was not directly involved in such attacks, or where the association with al-Qaeda was looser? Where should the legal line be drawn, and how can we in the public know that it is being drawn responsibly and enforced? Given the difficulty of such determinations and the legal stakes of getting it right in each case, I suggest that there is an urgent need for a credible, competent and independent oversight authority to assess the combat status of potential targets. One possible means of providing greater oversight, while simultaneously respecting the sensitivity of some of the intelligence relevant to making such determinations, would be the creation of independent and authoritative judicial bodies to review the combat status of potential targets *in camera.*[37]

In summary, an examination of the US targeting of Qaed Salim Sinan al-Harethi on 3 November 2002 suggests that a forceful (if not necessarily conclusive) legal justification could be offered for the operation. However, this same analysis raises doubts about whether this legal case can necessarily be generalized to cover all US targeting operations (or potential operations). In particular, it suggests the need for the US government to be clearer in articulating and defending what it believes to be the appropriate criteria for warranting such operations, and to build in more oversight and safeguards to assure that these criteria are strictly and rigorously adhered to.

Notes

[1] L. Calhoun, 'The Strange Case of Summary Execution by Predator Drone,' *Peace Review*, Vol. 15, No. 2, 2003, pp. 209-10.

[2] BBC News, 'US Still Opposes Targeted Killing,' November 6, 2002, Accessed 21 Jan. 2006, <http://news.bbc.co.uk/2/hi/middle_east/2408031>.

[3] J. Meyer, 'CIA Expands Use of Drones in Terror War,' *Los Angeles Times*, 29 Jan. 2006, Accessed 20 March 2007, <http://www.commondreams.org/headlines06/0129-05.htm>.

[4] United Nations Commission on Human Rights, 'Civil and Political Rights, Including the Questions of Disappearances and Summary Executions: Extrajudicial, Summary or Arbitrary Executions: Report of the Special Rapporteur Asma Jahangir, Submitted Pursuant to Commission on Human Rights Resolution 2002/36,' E/CN.4/2003/3, 13 January 2003, <http://www.extrajudicialexecutions.org/reports/E_CN_4_2003_3.pdf>, Paragraph 39; United Nations Commission on Human Rights, 'Civil and Political Rights, Including the Questions of Disappearances and Summary Executions: Extrajudicial, Summary or Arbitrary Executions: Report of the Special Rapporteur Philip Alston,' E/CN.4/2005/7, 22 December 2005, Accessed 5 February 2007 <http://www.extrajudicial executions.org/reports/E_CN_4_2005_7.pdf>, Paragraph 41.

[5] Major M. Machon, *Targeted Killing as an Element of US Foreign Policy in the War on Terror*, Monograph, School of Advanced Military Studies, Fort Leavenworth, Kansas, 2006, Accessed 4 February 2007, <http://www.fas.org/irp/eprint/machon.pdf>, pp. 2-3.

[6] Amnesty International, 'An Extrajudicial Execution by the CIA,' 18 May 2005, Accessed 2 February 2007, <http://web.amnesty.org/library/Index/ENGAMR510792005?open&of=ENG-384m.

[7] C. Gray, *International Law and the Use of Force, Second Edition*, Oxford University Press, New York, 2004, p. 175; C. Downes, 'Targeted Killing in an Age of Terror,' *Journal of Conflict and Security Law*, Vol. 9, No. 2, 2004, pp. 279-80.

[8] H. Duffy, *The 'War on Terror' and the Framework of International Law*, Cambridge University Press, Cambridge, 2005, pp. 342-344.

[9] United Nations Commission on Human Rights, 2005, op. cit., paragraphs 41 and 43.

[10] M. Schmitt, 'State-Sponsored Assassination in International and Domestic Law,' *Yale Journal of International Law*, Vol. 17, No. 1, p. 644; Y. Dinstein, *The Conduct of Hostilities Under the Laws of Armed Conflict,* Cambridge University Press, Cambridge, 2004, pp. 94-5; also see J. Kendall, 'Israeli Counter-Terrorism: 'Targeted Killings' Under International Law,' *North Carolina Law Review*, Vol. 80, No. 3, 2002.

[11] High Court of Justice, *Public Committee against Torture in Israel et al.* v. *Government of Israel et al.*, 14 December, 2006, Accessed 11 February 2007 <http://elyon1.court.gov.il/files_eng/02/690/007/a34/02007690.a34.pdf>.

[12] K. Roth, 'Debating the Issues,' *International Law Studies*, Vol. 81, 2006, p. 298.

[13] M. O'Connell, 'When is a War not a War? The Myth of the Global War on Terror,' *ILSA Journal of International and Comparative Law*, Vol. 12, No. 2,

2005, p. 5; D. Byman, 'Do Targeted Killings Work?' *Foreign Affairs*, Vol. 85, No. 2, 2006, p. 96; Downes, op. cit., p. 278.

[14] United Nations Commission on Human Rights, 2005, op. cit., paragraph 41.

[15] Downes, op. cit., pp. 280-1; United Nations Commission on Human Rights, 2003, op. cit., paragraph 37-8.

[16] S. Hersh, ''Manhunt': The Bush Administration's New Strategy in the War against Terrorism,' *The New Yorker,* 23 and 30 December, 2002, accessed 12 January 2007, <http://www. newyorker.com/fact/content/articles/ 021223fa_fact>.

[17] United Nations Commission on Human Rights, 2003, op. cit., paragraph 38.

[18] B. Whitaker, 'Cornered Yemeni al-Qaida Suspect Blows Himself Up,' *Guardian Unlimited,* 14 February, 2002, Accessed 12 February 2006, <http://www.guardian.co.uk/yemen/ Story/0,2763,649933,00.html>; also see United Nations Commission on Human Rights, 2003, op. cit., paragraph 38, for a report of several unsuccessful efforts to arrest al-Harethi.

[19] W. Banks, 'Legal Sanctuaries and Predator Strikes in the War on Terror,' in Michael Innes (ed.), *Denial of Sanctuary*, Praeger Security International, London, 2007, p. 117.

[20] Downes, op. cit., p. 281; Duffy, op. cit., p. 337.

[21] Downes, op. cit., p. 280; Y. Dinstein, *War, Aggression and Self-Defence*, Cambridge University Press, New York, 2007, pp. 112-6.

[22] Downes, op. cit., p. 281; Inter-American Commission on Human Rights, *Report on Terrorism and Human Rights*, General Secretariat, Organization of American States, Washington, 2002, paragraph 87.

[23] M. Schmitt, 'Counter-Terrorism and the Use of Force in International Law,' *International Law Studies*, Vol. 79, 2003, p. 11; Dinstein, *War, Aggression and Self-Defence*, op. cit., pp. 206-207.

[24] International Court of Justice, *Legality of the Threat or Use of Nuclear Weapons*, 8 July 1996, Accessed 21 January 2007, <http://www.icj-cij.org/docket/files/95/7495.pdf>, p. 245; Dinstein, *War, Aggression and Self-Defence*, op. cit., p. 208.

[25] Dinstein, *War, Aggression and Self-Defence*, op. cit., p. 250; Schmitt, 'Counter-Terrorism and the Use of Force in International Law,' op. cit., p. 36.

[26] Dinstein, *War, Aggression and Self-Defence*, op. cit., p. 250.

[27] Gray, op. cit., p. 124.

[28] Dinstein, *War, Aggression and Self-Defence*, op. cit., p. 210; M. O'Connell, 'The Myth of Pre-emptive Self-Defence,' *American Society for*

International Law Task Force on Terrorism, August 2002, Accessed 10 November 2007, <http://www.asil.org/taskforce/oconnell.pdf>, p. 9.
[29] Schmitt, 'State-Sponsored Assassination,' op. cit., p. 649.
[30] Schmitt, 'State-Sponsored Assassination,' op. cit., p. 648.
[31] S. Carapico, 'Yemen and the Aden-Abyan Islamic Army,' *Middle-East Report Online*, 18 October 2000, Accessed 27 June 2008, <http://www.merip.org/mero/mero101800.html>.
[32] Duffy, op. cit., p. 252.
[33] Dinstein, *The Conduct of Hostilities*, op. cit., p. 94.
[34] S. Ratner, 'Predator and Prey: Seizing and Killing Suspected Terrorists Abroad,' *the Journal of Political Philosophy*, Vol. 15, No. 3, 2007, p. 275.
[35] Roth, op. cit., pp. 398-399.
[36] United Nations Commission on Human Rights, 2005, op. cit., paragraph 41.
[37] A. Plaw, 'Terminating Terror,' *Theoria*, Vol. 114, 2007, pp. 23-25.

Bibliography

Amnesty International, 'An Extrajudicial Execution by the CIA'. 18 May, 2005, Accessed 2 February 2007, <http://web.amnesty.org/library/Index/ENGAMR510792005?open&of=ENG-384m>.

Banks, W., 'Legal Sanctuaries and Predator Strikes in the War on Terror.' *Denial of Sanctuary*. Innes, M. (ed), Praeger Security International, London, 2007.

BBC News, 'US 'Still Opposes' Targeted Killing.' November 6, 2002, Accessed 21 January 2006, <http://news.bbc.co.uk/2/hi/middle_east/2408031.stm>.

Byman, D., 'Do Targeted Killings Work?' *Foreign Affairs*. Vol. 85, No. 2, 2006, pp. 95-112.

Calhoun, L., 'The Strange Case of Summary Execution by Predator Drone.' *Peace Review*. Vol. 15, No. 2, 2003, pp. 209-214.

Carapico, S., 'Yemen and the Aden-Abyan Islamic Army.' *Middle-East Report Online*. 18 October 2000, Accessed 27 June 2008, <http://www.merip.org/mero/mero101800.html>.

Dinstein, Y., *War, Aggression and Self-Defence.* Cambridge University Press, New York, 2007.

—, *The Conduct of Hostilities under the Laws of Armed Conflict.* Cambridge University Press, Cambridge, 2004.

Downes, C., 'Targeted Killing in an Age of Terror.' *Journal of Conflict and Security Law.* Vol. 9, No. 2, 2004, pp. 277-294.

Duffy, H., *The War on Terror and the Framework of International Law.* Cambridge University Press, Cambridge, 2005.

Gray, C., *International Law and the Use of Force, Second Edition.* Oxford University Press, New York, 2004.

Hersh, S., 'Manhunt: The Bush Administration's New Strategy in the War against Terrorism.' *The New Yorker.* 23 and 30 December 2002, accessed 12 January 2007, <http://www. newyorker.com/fact/content/articles/021230>.

High Court of Justice, *Public Committee Against Torture in Israel et al.* v. *Government of Israel et al.* 14 December 2006, Accessed 11 February 2007, <http://elyon1.court.gov.il/files_eng/02/690/007/a34/02007690.a34.pdf>.

Inter-American Commission on Human Rights, *Report on Terrorism and Human Rights.* General Secretariat, Organization of American States, Washington, 2002.

International Court of Justice, *Legality of the Threat or Use of Nuclear Weapons*, 8 July 1996, Accessed 21 January 2007, <http://www.icj-cij.org/docket/files/95/7495.pdf>.

Kendall, J., 'Israeli Counter-Terrorism: 'Targeted Killings' under International Law.' *North Carolina Law Review.* Vol. 80, No. 3, 2002, pp. 1069-1088.

Machon, M., *Targeted Killing as an Element of US Foreign Policy in the War on Terror.* Monograph, School of Advanced Military Studies, Fort Leavenworth, Kansas, 2006 Accessed 4 February 2007, <http://www.fas.org/irp/eprint/machon.pdf>.

Meyer, J., 'CIA Expands Use of Drones in Terror War.' *Los Angeles Times.* 29 January 2006, Accessed 20 March 2007, <http://www.commondreams.org/headlines06/0129-05.htm>.

O'Connell, M., 'When is a War not a War? The Myth of the Global War on Terror.' *ILSA Journal of International and Comparative Law.* Vol. 12, No. 2, 2005, pp. 1-5.

—, 'The Myth of Pre-emptive Self-Defence.' *American Society for International Law Task Force on Terrorism.* August, 2002, Accessed 10 November 2007, <http://www.asil.org/taskforce/oconnell.pdf>.

Plaw, A., 'Terminating Terror.' *Theoria.* Vol. 114, 2007, pp. 1-27.

Ratner, S., 'Predator and Prey: Seizing and Killing Suspected Terrorists Abroad.' *The Journal of Political Philosophy.* Vol. 15, No. 3, 2007, pp. 251-275.

Roth, K., 'Debating the Issues.' *International Law Studies.* Vol. 81, 2006, pp. 396-416.

Schmitt, M., 'Counter-Terrorism and the Use of Force in International Law.' *International Law Studies.* Vol. 79, 2003, pp. 7-73.

—, 'State-Sponsored Assassination in International and Domestic Law.' *Yale Journal of International Law.* Vol. 17, No. 1, 1992, pp. 609-685.

United Nations Commission on Human Rights, 'Civil and Political Rights, Including the Questions of Disappearances and Summary Executions: Extrajudicial, Summary or Arbitrary Executions: Report of the Special Rapporteur Asma Jahangir, Submitted Pursuant to Commission on Human Rights Resolution 2002/36.' E/CN.4/2003/3, 13 January 2003, Accessed 5 February 2007, <http://www.extrajudicialexecutions.org/reports/E_CN_4_2003_3.pdf>.

United Nations Commission on Human Rights, 'Civil and Political Rights, Including the Questions of Disappearances and Summary Executions: Extrajudicial, Summary and Arbitrary Executions: Report of the Special Rapporteur Philip Alston.' E/CN.4/2005/7, 22 December, 2005, Accessed 5 February 2007, <http://www.extrajudicialexecutions.org/reports/E_CN_4_2005_7.pdf>.

Whitaker, B., 'Cornered Yemeni al-Qaida Suspect Blows Himself Up.' *Guardian Unlimited.* 14 February, 2002, Accessed 12 February 2006, <http://www.guardian.co.uk/yemen/ Story/0,2763,649933,00.html>.

Avery Plaw is an Associate Professor of Political Science at the University of Massachusetts at Dartmouth. He is the author of *Targeting Terrorists: A License to Kill?*, and *Frontiers of Diversity: Explorations in Contemporary Pluralism.*

Insecurity by Impreciseness: Towards a Specific Concept of Security

Benjamin Rampp

Abstract

The aim of the following exposition is to critically analyse the concept of human security in security studies and suggest some methodological refinements. First, I will comment on the gradual expansion of the classic concept of security. This expansion comprises two dimensions—deepening and broadening. I will then illustrate how the expansion of security in general is reproduced in the concept of human security and why this presents a problem. To this end, I will first introduce the human security approach and then elaborate on the advantages and disadvantages of the expansion in this context. In particular, I will show how the broadening of human security and its resultant conceptual ambiguity prove to be a fundamental problem. I will establish that the conceptual impreciseness has negative and—regarding human security's original intention—counterproductive impacts on the concept's political effect. Here the securitization of originally non-security issues will be central to my argument. This problem with human security is not purely of academic interest, but has impacts on practice, as is illustrated and discussed at the end of this chapter in relation to humanitarian intervention. As a preliminary solution I suggest (from a methodological point of view) the use of a 'core concept' of human security which concentrates on the physical inviolability of the individual. Finally, I will elucidate this narrow conception as a first step towards a broader desecuritization process aimed at preserving the adequateness of security ascriptions.

Key Words: Security, human security, redefined security, securitization, desecuritization, Copenhagen School, state of exception, methodology, constructivism.

*****

1. Expansion of the Concept of Security and Human Security

A. Expansion: Deepening and Broadening

The classic concept of (national) security, which primarily references (1) the state as central actor and (2) physical-military dangers, has been questioned for a fairly long time. In contrast to its limited focus, an expansion of the concept has been suggested which applies to two dimensions. The first dimension can be termed 'deepening.'[1] It describes a shift of the level of analysis on the vertical axis (from the state downwards to the individual).[2] The second dimension denotes expanded fields of application of the security concept by 'broadening.'[3] Here, security is expanded on the horizontal axis and comprises issues over and above physical inviolability.[4]

According to Roland Paris' classification on both axes (horizontal: 'What is the Source of the Security Threat?'—classifications 'Military' and 'Military, Nonmilitary, or Both;' vertical: 'Security for Whom?'—classifications 'States' and 'Societies, Groups and Individuals') four fields of security studies can be distinguished: national security (Military/States), redefined security (Military, Nonmilitary, or Both/States), intrastate security (Military/Societies, Groups and Individuals), and human security (Military, Nonmilitary, or Both/Societies, Groups and Individuals).[5] In the field of national security, the state's territorial integrity is central. Redefined security still focuses on the state, but the threats to its security are derived from a broader spectrum of sources (including ecological, economic, intrastate and sub-state threats like terrorists). Intrastate security denotes military conflicts within the state, such as civil wars. Finally, human security (in this scheme) has undergone deepening and broadening expansion and thus relates to both groups and individuals as well as to a broad spectrum of threats.[6]

This expansion of the concept of security occurs in the context of a change of paradigm in which it becomes possible to speak not only about clearly attributable dangers or threats but also about *risks*.[7] These are ubiquitous and to some degree dependant on one's own actions. Thus, security is no longer a question of specific attributions and polarities, but is changing continuously and has to be guaranteed constantly. For this reason, each actor and each action becomes a security issue. Security is therefore constantly in danger—the state of exception becomes the rule.[8]

The two dimensions of expansion have been confronted by diverse criticism, some—although not all—emanating from supporters of the classic notion of security. The process of deepening the concept of security is disputed because it is no longer the state but individuals who are made the relevant actors and addressees of security studies. In this, the classic agent-structure debate appears: is the individual actor central or is the actor mainly influenced by the surrounding structure (or vice versa)? There are further objections to the deepening of the concept of security which criticise its individualization as being overly accentuated.[9] The broadening of security is also subject to criticism. It is argued, for example, that the expansion of the spectrum of threats and risks distracts from relevant dangers and excessively dilutes the concept of security.

B. Human Security

This phenomenon of conceptual expansion also provides the context for the development of the notion of human security. Human security is a topic which has constantly been discussed by this name in the context of international relations and human rights since, at the very latest, 1994 (and the publication of 'Human Development Report 1994' by the United Nations Development Programme).[10] However, the basic underlying thought is much older and is found, for example, in the term αταραξία (ataraxia), which implies a holistic idea of security with the additional connotation of 'peace of mind.' Indeed, ataraxia anticipates the allegedly 'new' concept of human security—with all its pros and cons. There have been long discussions and animadversion concerning the usefulness and side effects of the concept. These discussions continue with renewed intensity today and many questions are unresolved.[11] The concept is, however, still used frequently in contemporary discourse and in its broadest interpretation. The critique too persists.

Human security is based on two ideas, which respectively reflect the aforementioned dimensions of the expansion of the classic security concept: (1) that the state should no longer be the referential object of security studies, but social groups and the individual itself (deepening), and (2) that securing only the immediate physical inviolability is insufficient. Rather, structural violence[12] should be a major focus—that is, indirect violence and the causes of, for example, famine, environmental pollution, inadequate education, insufficient political rights and lack of dignity and development. Human security then denotes a more comprehensive concept in respect to this continuum of threats (broadening). Depending on the extent of expansion of the classic national security concept, a narrow and a broad approach of human security can be distinguished. While both notions support deepening the classic concept of security, they differ in regard to broadening: the narrow approach conceives human security mainly as *freedom from fear* and especially addresses security against violent physical threats; the broad approach additionally interprets human security as *freedom from want* and thus also includes security against socio-economic threats (famine, environmental pollution, inadequate education, insufficient political rights and lack of dignity).[13] In this light, the broad approach appears as a holistic concept which is more concerned with human development in general than strictly with security.[14] In recent years, the broad concept of human security has gained increasing traction in public discussion. A question then has to be asked: how much does such a comprehensive human security approach specifically contribute to solving (political) security challenges, or might it not stretch the concept of security too far?

2. A Criticism of the Human Security Approach

Criticism of the expansion of the concept of security applies especially to the notion of human security—particularly in its broad variant—because a maximum of increased conceptual flexibility can be found here.

A. Criticism of Deepening

Critics of deepening the concept of security—that is, critics of drawing social groups and individuals into focus instead of the state—point out that states are still the central element of the political. Certainly, they acknowledge, other actors have also gained access to the political arena, however the state is still the foundation of political action. This criticism is raised by supporters of the classic concept of security as well as by some supporters of Critical Security Studies: for example, Mohammed Ayoob stresses that the majority of worldwide conflicts take place in the third world, are domestic and depend on the success/failure of the state building process.[15] Ayoob concludes that only an existing state is able to guarantee its inhabitants security. That is, individual security is supposed to be primarily a question of the state, and can—at least in the first step—only be guaranteed by it. In other words, the state is the necessary condition of security, on the basis of which considerations about individual security can be made, but without which these considerations are doomed to fail from the start. In this respect, deepening the concept of security distracts from the world's 'real' security problems.

I partly concur with this criticism. Of course, the state is normally some guarantee of security and in most cases, some state is better than none for people's security. However, this is certainly not true for all states, especially when security is conceived in a wider sense. There are numerous examples of how states structurally create insecurity to realise a 'governmentality of unease'—as Didier Bigo, for example, demonstrates with the example of normalisation politics towards immigrants.[16] Nonetheless, states are prima facie, all exceptions considered, necessary—albeit *not sufficient*—conditions for the guarantee of minimal physical security. So the question has to be asked whether this argues against the expansion of the concept of security to include social groups and individuals. Yet just because the focus changes, the state as analytical category doesn't have to vanish. On the contrary, the individual's security must be considered specifically by itself as well as in the state's context (as the state undoubtedly belongs to the individual's social reality). It would certainly be absurd to limit the focus of security to the individual and thereby ignore the state—something no one wants to do anyway. Rather, the claim on which deepening is based is simply the need to move beyond an *exclusive* focus on the state—an exclusivity characteristic of many neorealists' models.[17] The individual as category should be involved to elaborate on the interconnected processes happening in the real world. In doing so, the respective levels of analysis must remain conscious and must not be amalgamated. Framed this way, I am persuaded of the value of expanding the focus of security beyond the state towards the individual. This is not because I prefer the individual over the state in relation to security in general, but simply because the individual should not, in principle, be removed from the field of view. Rather, from a pragmatic point of view, different empirically-induced approaches should be available. If *human* security is thus discussed, security should somehow address individual *humans*. One attractive framework for integrating individual and institutional elements is Anthony Giddens' mediating approach, which theorizes the mutual structuration of agent and structure, and thus leaves all possibilities available.[18] Another worthwhile alternative would of course be to leave the artificial separation of agent and structure behind and to concentrate on practice (i.e., the social process) rather than emphasizing the two poles[19]—an approach that Didier Bigo supports in the field of securitization (see below).

B. Criticism of Broadening

While I think of deepening, when used consciously and specifically, as a worthwhile dimension of expansion, broadening seems to be more problematic. The weaknesses of the human security approach caused by broadening can be seen (in a paradigmatic broad approach) in the characterisation of human security in the Human Development Report. It states that 'Human security is not a concern with weapons—it is a concern with human life and dignity.'[20] The report describes seven areas of human security: economic, food, health, environmental, personal, community and political security.[21] This description summarises a rather large continuum ranging from physical violence to human dignity. The addressed topics show that this broader version deals less with a specific concept of security than with a more general understanding of development and (psychosocial[22]) well-being.

The question then arises, why is the area still framed as security and what value does such an all encompassing concept offer? If a theory or a concept incorporates everything, it becomes a non-falsifiable metaphysics—unusable as a scientific instrument.[23] Paris commented on this with the well-known phrase: 'Again, if human security is all these things, what is it not?'[24] Müller follows a similar reasoning (originally regarding the redefined security concept) by stating that 'life itself is life threatening.'[25] He argues that it doesn't make sense to label everything as 'security' and thus void the term itself, and respectively redefine and re-evaluate other topics. If security is an attributive-only characteristic, which can be randomly attached to other fields, it becomes meaningless. Even worse, analytical categories—which have an impact on practice—are blurred. The value of attributing anything at all to security remains unclear.[26]

The dilution of security and the problems associated with it are not only an issue of some single case, or abstract model. Instead there is a structural connection between the expansion of the concept of security and its frequently negative impact in the real world. This structural interrelationship can be illustrated more generally with the concept of 'securitization.' In this approach, security is not conceptualised as a real world subject-matter, but rather it is taken as a premise that issues are actually made subjects of security. There are different securitization approaches within security studies, all of which take the production of security and security issues as a starting point, but all of which also emphasise varying elements of the process.[27] The 'Copenhagen School', for example, conceptualises security as a speech act—that is, the term 'security' is assigned (by power holders) to issues and therefore is realised.[28] The so-called Critical Security Studies are similar to this approach, but especially emphasise a normative take on the field of security studies and the necessity of the emancipation of the individual from the state as a reference point for security considerations. Finally, the 'Paris School' with Didier Bigo as a main influence emphasises—following Bourdieu (and partly Foucault)—the relevance of the practice (and not only the speech acts) of securitizing actors. It empirically investigates the respective institutions' fights over power and interpretive authority in the field of security. All differences aside, the approaches have in common that they treat security not as a noumenon, but as a socially constructed issue. This constructed character of the subject allows discussion and expansion of the issues and reference object of security—as can be seen in the case of the concept of risk. The securitization approaches show furthermore, that the expansion's quality is not derived from the topic itself, but depends respectively on discursive (Wæver) and non-discursive (Bigo) practices of the securitizing actors. Securitization is thus the result of social action and therefore not a given fact. With securitization, aims formulated by securitizing actors—or conglomerates of securitizing actors—are pursued (where the 'common' aims of these conglomerates are not necessarily pursued collectively on purpose, but may as well be the results of the reciprocity of non-intended effects of different social actions). Accordingly, there is also the possibility, and perhaps the necessity, of desecuritization.

Securitization basically describes a process in which originally differentiated fields (economy, food, health, environment, society, etc.) and political issues are associated with the term and therefore with the logic of security—so that they become questions of security and thus their borderlines are actively blurred. By consequence, each respective area achieves—which in some cases might be useful and necessary—more attention on the one hand ('It's the security, stupid!'), but also becomes part of another 'game' on the other. Ole Wæver writes—similar to Paris—about the basic problem, which especially relates to the human security approach:

> The security of individuals can be affected in numerous ways; indeed, economic welfare, environmental concerns, cultural identity, and political rights are germane more often than military issues in this respect. The major problem with such an approach is deciding where to stop, since the concept of security otherwise becomes a synonym for everything that is politically good or desirable. How, then, can we get any clear sense of the specific character of *security* issues, as distinct from other problems that beset the human condition?[29]

In this context, security becomes the ultimate principle of the social world.[30] Thus the logic of security is extended to other social fields, which however lose specificity.[31] If, for example, questions of human development or the development of democracy are seen through the eyes of security, then these questions will also be answered from security's point of view (and not in respect to their original aims). This evolution can be observed on all levels: instead of focusing, for example, on the long term development of democratic institutions and culture, a system might be—in the name of (human) security—overthrown with military means to then try reconstruction from the left over ruins. Alternatively, regarding the domestic sphere, instead of supporting begging persons to sustainably improve their situation, their conduct is forbidden in the name of public security—and the respective individuals therewith driven away from commercial areas where they are regarded as disturbing elements.[32] Essentially, different situations are thus treated in a uniform way that might at times be counterproductive. Wæver expresses this even more drastically:

> Consequently, the logic of war—of challenge-resistance (defense)-escalation-recognition/defeat—could be replayed metaphorically and extended to other sectors. When this happens, however, the structure of the game is still derived from the most classical of classical cases: war.[33]

The question arises whether such a logic of security—which in practice comprises more terms like intentional threats, us vs. them,[34] coercion, violence and counter violence, and the loss of freedoms rather than the idea of 'peace of mind'—is appropriate for the respective situations. I strongly doubt this. Rather, I fear that potentially better alternative approaches to solve problems in the concrete situation are ignored by securitizing them.[35]

C. Suggestion: Deepening, not Broadening

My basic argument concerning the expansion of the concept of security using the example of human security can be summarized as follows: because of the risks of conceptual arbitrariness and inappropriate securitization, broadening the concept of security should be avoided. Security should be treated as an exclusive category which allows discovering interconnections between relevant variables. By contrast, I think that deepening, when applied thoughtfully, is worthwhile, because human security should reflect the individual level—the *human* in human security—and contains more than the classic national security concept. Obviously, there is a danger of amalgamation of categories and securitization in both dimensions of expansion. But I assume that inappropriate securitization can be avoided more easily in deepening than in broadening because levels of analysis can be better distinguished and separated after deepening as compared to the contents which mingle and become inseparable after broadening.

A central characteristic of a meaningful and practicable narrow human security approach seems therefore to deepen the classic (national) security concept but not to broaden it. In the centre of such a narrow human security approach, which is in some way similar to the freedom from fear variant of human security, there would be a concept with a focus on 'personal security'[36]—that is, a 'core concept' dealing with the basic idea of the physical inviolability of individuals as a common denominator.

D. A Question of Methodology

It must be emphasised that my argument is not about the ontological question whether security and issues of development/well-being (economy, food, health, environment, social dimensions) are interconnected or not—of course these issues are related: the social world is a complex and multicausal one. I'm much more concerned with the epistemological and methodological question of how these relations can be investigated. To avoid an opaque amalgamation of topics, I therefore suggest—from a methodological point of view—that we first describe fields of practice exclusively and only then investigate their connections in the real world and reintegrate them. With the conclusion that everything is interrelated (or even is one) nothing is gained. The question is '*how* is everything interrelated?'

I'm not proposing a scientific realist approach. Whether security, development and well-being are different subjects in reality or not is of no interest here. But to work with them in practice, the terms have to be defined. These definitions are not real in the sense of a naturalistic understanding (and should therefore perhaps better be named descriptions).[37] But they *become* real by using them. Therefore, it is better to describe them deliberately and explicitly than to accept implicit definitions. Hence, the claim for a narrow approach is not derived from a traditional realist (not in the sense of scientific realism, but in the tradition of International Relations) approach,[38] but rather can be based on a constructivist premise.

The explicit specification of the concept of security, less along the lines of a comprehensive human security approach and more like a concept focusing on direct physical inviolability of the individual, appears initially as a step back. This step, however, is necessary for the causes of (in-)security—the explanans—to no longer be part of the description of (human) security—the explanandum. Only by redefining security as security, and economy, food, health, environment and the social dimensions as specific terms respectively, does it become possible to differentiate the issues theoretically and thereby uncover relations between causes and effects (correlations, possibly even causalities?).[39] Thus, security—as attributive-only characteristic without exclusive value and just as a context variable—has to be returned to a variable with genuine intrinsic value.[40]

3. Not an Academic Discussion, but one with Real Consequences

The criticism of expanding the concept of security to form a broad human security approach is not purely academic, but concerns reality. Human security as an analytical concept is simply too vague to serve as a political category offering orientation for actions. The concept may be of interest as a general normative-ethical position, but is not applicable in any more concrete way.[41] Moreover, this vagueness is dangerous—because of its lack of analytical preciseness and the consequent broadening of the reference spectrum. As a result, the human security approach may even result in opposite outcomes to its actual goals, as was demonstrated with the process of securitization. Theory and practice cannot be separated, because the theoretical formulation of human security is part of the practice of securitization. In that light, it seems problematic to criticize the analytical dimension of the concept of human security and instead to emphasize its political-normative utility, as proposed for example by Rita Floyd.[42]

An example of the potentially counterproductive effects of too excessive a conceptual expansion is humanitarian intervention and therefore the question of just war, which again became current with the success of the human security approach.[43] I don't want to reject humanitarian intervention in principle as inappropriate securitization. Instead, the question has to be asked, when is such an instrument 'inappropriate?' Here, my assessment of the two dimensions of expansion again comes into play. I think it is worthwhile to use the concept of security for individuals—even when, by doing this, the sovereignty of the state might be weakened and intrusions into 'domestic' affairs of the state are legitimated. However, this has to happen under very special circumstances. The result of intervention must not be worse for the individual than if there were no intervention. Judging this—particularly à la longue—will always be problematic in practice. The risk remains that harming of a (broadly conceptualised) human security will be abused as an arbitrary reason for warfare and that public international law will thereby be sacrificed in the name of human rights—which does not appear to be reasonable. Public international law is still an important stabilising instrument, although it must not be the solitary dominating category of political action, which is also true of (individual) human rights; therefore, a pragmatic interaction between these two dimensions is necessary, not a dogmatic confrontation. By conceptualising human security in the horizontal dimension as narrowly as possible, the danger of abuse is minimised.

The foundation of an appropriate military humanitarian intervention is therefore in my opinion to focus on the individual (deepening), but only from a topically-narrow, not broadened, perspective so that not everything turns into a question of humanitarian intervention, but only cases of structural, mass and severe harm of the universal right to life and the physical inviolability of the individual (which might well be impossible and undesirable to ever operationalise absolutely). Humanitarian intervention has to be conducted as ultima ratio,[44] within specified[45] and sufficiently promising terms, as well as with clearly defined targets and exit conditions—but not as a standard answer to basic dangers to human life or value conflicts in general. In some ways, I follow Walzer here, when he emphasises that there are certain qualitative limits, the transgression of which might require measures of violence and at the same time tries to delimit the occasion of such a reaction as narrowly as possible: 'We are best served, I think, by a stark and minimalist version of human rights here: it is life and liberty that are at stake.'[46] These conditions however do not apply in case of a (horizontally) broadened human security as the basis of a humanitarian intervention. In contrast to the already quoted statement 'Human security is not a concern with weapons—it is a concern with human life and dignity,'[47] this possibility shows that human security can very well develop into such a question, in which the imprecision of the concept supplies possibilities for military actions. For instance, after the end of the Cold War, not only did the world's security situation change fundamentally, but also increasingly often the rationale of waging war—war in the name of human security. The most current example for this is probably 'Operation Iraqi Freedom.' Whatever the reasons for which this war has been conducted, one of the arguments—moreover the longest lasting one—was the implementation of democracy in Iraq. This has shown that carrying human security on the flag doesn't mean actually achieving this goal—quite the contrary, as the International Committee of the Red Cross' report on Iraq war states.[48]

If broadened human security is thus used as a concrete political goal and guideline for actions, this might ironically lead to more insecurity rather than to more security. Therefore, a two-step procedure is necessary: by striving to achieve specific and clearly formulated security goals at first, there is a chance that, in a second step, human security in its most fundamental sense can be established.

4. Conclusions: A Starting Point for Desecuritization

We have seen several problems which are associated with the expansion of security to a broad concept of human security, and that a narrow concept of human security might be more reasonable. As a first step towards such a concept, I have suggested allowing deepening but rejecting broadening, and supporting a 'core concept' of human security based on the physical inviolability of individuals. This strategy is aimed at preventing inappropriate securitization getting out of hand. In the foregoing references to desecuritization, I have already mentioned the possibility to avoid and reverse securitization processes. When the political in a Schmittian sense is conceived as a fight between friend and enemy and therefore as a fight over interpretive dominance,[49] then it is also possible to again change interpretations. And this is not a mere question of power, but also a normative question. Although approaches of securitization, especially Wæver's,[50] are also based on Schmitt's 'constructivist' basis—that the political is not conceivable in an objectivist manner, but is a product of social action—they draw different conclusions from Schmitt. While he favours a legal positivism in a relativistic style in which anything that is lawful (i.e., that has prevailed) is right and thus everything can be right in a state of emergency, securitzation approaches emphasise the necessity of desecuritization, which implies that not every securitization is reasonable and that therefore there has to be an opposing process.

At this point, the first question is what is the 'right' level of securitization? (Since the absolute non-existence of securitization is pointless once the objectivistic character of security is abandoned.) The second question is how can this level be (re)installed?

First, as shown above using the example of humanitarian intervention, the line between appropriate and inappropriate securitization is thin. Security is produced and the (normative) limitation of production is not inherent to the object, but has to be discussed and decided on. Frequently, such a limitation can only be decided on *ex post*. Nevertheless, there are orientation aids, which can be applied in security politics. As was mentioned in the case of humanitarian intervention, a teleological argumentation can be pursued, that securitization has to lead to more and not to less security. The consequences of securitization, however, can only be assessed based on earlier experiences, which might not necessarily apply to the respective cases currently subject to investigation. The normative nature of the question of the right level of securitization eventually demands that close attention be paid to the generation process of securitization (i.e., one should begin with the actors who discuss the question of security, with the *process* of securitization itself). Securitization and desecuritization should happen in an open, public, democratic process, which—and for this the concept might be fundamentally useful—is based on ideas that are at the foundation of the notion of human security. Rules and limits which will prevent abuse of the idea of security will have to be consciously developed. In practise, this is hard to realise—because security questions frequently have to be treated discreetly and secretly. Nevertheless, designing foundations and regulations for the respective negotiation processes is possible and reasonable. It is therefore necessary to develop 'crisis-proof' security ethics (i.e., rules which are not immediately abandoned in the 'fog of emergency,' but which are enduring). The suggested limitation on the concept of security in regard to its reference objects (i.e., in regard to broadening) is a first basis for such rules, which are able to offer orientation for actions.

Second, to reinstall the 'right' level of securitization, a process of desecuritization might be implemented. What could this process look like when the limits of securitization are eventually established? The theoretical dimensions of desecuritization have been discussed before and I don't want to repeat them here, but will point out a few issues.[51] To desecuritize an issue, there are a range of possibilities. One can apply an argumentative approach (given that there are 'real' security issues as such) and can reason that the respective issue was not a 'real' security issue; one can also, however, try to understand the construction process of the security issue to correct it; finally, a counter-narrative can be developed which investigates the issue from another point of view.[52] The practice of desecuritization can therefore take a range of forms, starting with counter argumentation against attribution of issues (as is done in the case of the present chapter), up to consciously ignoring and omitting the concept of security in some contexts. In this context it has to be considered that desecuritization takes place in a social field of conflict in which not only the best argument wins, but also where actors' interests and the capabilities of enforcement help to decide which issues are securitized and which are not. It must also not be forgotten that there will also be resistance against efforts at descuritization. Hence, desecuritization is as much a question of the capability of enforcement and the endurance of the actors' engagement as it is a question of better arguments.

Of central relevance for the success of a desecuritization process might also be its point in time. Firstly, it has to be recognized that when a process of securitization has institutionalized itself and the attribution becomes a 'thickened' and 'hardened' objectivity,[53] when therefore issues have been transformed into questions of security in the individual and public consciousness, it is hard to reverse this process. Calling it into question early therefore seems to have better chances for success. Secondly, desecuritization does not happen in empty social space, just as little as securitization does. Events have influence on desecuritization's likelihood of success; (de)securitization is no detached discursive reality, but is connected with events of first order. Likeliness of success is therefore not only dependent on the point of time within the securitization process, but also on context (e.g., high publicity acts of terror obviously facilitate securitization and make desecuritization difficult). Desecuritizing should therefore be easier in calm and peaceful times. Still, this assumption has to be restricted because the more social reality is interpreted in the paradigm of risk, the harder it will be to find conditions which are *not* issues of security. In this light, the likely strategies of securitizing actors—which include proclaiming and exploiting (permanent and comprehensive) states of emergency—are still important for the development of counter strategies. The focus of desecuritization therefore has to be put first on preventing the emergence of an omnipresent state of emergency in order to eventually have an opportunity to handle concrete cases at all. In the end, the central concern of all of the different approaches to desecuritization is to avoid Schmittian thinking patterns of the political, which regard politics as a confrontation between friend and enemy.[54] Finally, nowhere will this work better than in an open and reflective discourse in which the role of security studies is itself a topic of discussion.

Notes

I would like to thank Michael Nagenborg, Ari Ofengenden, Regina Ammicht Quinn and Stefan Rampp for their enriching comments and help. I would also like to thank the editor for thoroughly reviewing and thereby improving this chapter.

[1] R. Paris, 'Human Security: Paradigm Shift or Hot Air?' *International Security*, Vol. 26, Fall 2001, p. 97.
[2] E. Rothschild, 'What is Security?' *Daedalus*, Vol. 124, Summer 1995, p. 55.
[3] Paris, op. cit., p. 97.
[4] Stuart Murray's contribution in this book addresses a similar case of concept expansion—the expansion of the understanding of diplomacy. In contrast to my following argumentation versus an extensive expansion of the concept of security (and especially broadening), he supports a positive position in regard to the expansion of the concept of diplomacy. Although I am sceptical about holistic, all-encompassing concepts and their

practicability, I don't see as severe a danger in the expansion of diplomacy concepts as in the field of security, because a 'diplomatization' of politics (to a certain limit) seems more useful and less dangerous to me than securitization (more in the following).

[5] Paris, op. cit, pp. 97-100.

[6] C. Ulbert, *Human Security—ein brauchbares Konzept für eine geschlechtergerechte außen—und sicherheitspolitische Strategie?* Vortrag auf der Fachtagung des Feministischen Instituts der Heinrich-Böll-Stiftung in Kooperation mit der Friedrich-Ebert-Stiftung und dem Frauensicherheitsrat, Abgeordnetenhaus Berlin, Berlin 24/25, October 2003, Viewed on 24 November 2011, <http://www.glow-boell.de/media/de/txt_rubrik_3/Ulbert_autorisiert.pdf>.

[7] C. Daase, 'Internationale Risikopolitik. Ein Forschungsprogramm für den sicherheitspolitischen Paradigmenwechsel,' *Internationale Risikopolitik. Der Umgang mit neuen Gefahren in den internationalen Beziehungen*, C. Daase, S. Feske and I. Peters (eds), Nomos, Baden-Baden, 2002, pp. 9-35; see also U. Beck, *Risk Society: Towards a New Modernity*, Sage, New Delhi, 1992; U. Beck, *World Risk Society*, Polity Press, Cambridge, 1999; U. Beck, *Weltrisikogesellschaft. Auf der Suche nach der verlorenen Sicherheit*, Suhrkamp, Frankfurt am Main, 2007.

[8] G. Agamben, *State of Exception*, University of Chicago Press, Chicago, 2005.

[9] B. Buzan & O. Wæver, 'Slippery? Contradictory? Sociologically Untenable? The Copenhagen School Replies,' *Review of International Studies*, Vol. 23, April 1997, p. 245.

[10] United Nations Development Programme (UNDP), *Human Development Report 1994*, UNDP, New York, 1994, Viewed on 24 November 2011, <http://hdr.undp.org/en/reports/global/hdr1994/chapters/>.

[11] P.J. Burgess & T. Owen (eds), 'Special Section: What is 'Human Security'?' *Security Dialogue*, Vol. 35, September 2004, pp. 345-387.

[12] J. Galtung, 'Violence, Peace, and Peace Research,' *Journal of Peace Research*, Vol. 6, March 1969, pp. 167-191.

[13] G. King & C.J.L. Murray, 'Rethinking Human Security,' *Political Science Quarterly*, Vol. 116, Winter 2001-2002, p. 590; F.D. Roosevelt, *The Annual Message to the Congress*, 6 January 1941, Viewed on 24 November 2011, <http://www.presidency.ucsb.edu/index.php?pid=16092#axzz1eeXELadb>.

[14] T. Debiel & S. Werthes (eds), *Human Security on Foreign Policy Agendas. Changes, Concepts and Cases*, INEF Report 80/2006, Institute for Development and Peace, University of Duisburg-Essen, Duisburg, 2006, Viewed on 23 November 2007, <http://inef.uni-due.de/page/documents/Report80.pdf>, p. 10.

[15] M. Ayoob, 'Defining Security: A Subaltern Realist Perspective,' *Critical Security Studies*, K. Krause & M.C. Williams (eds), University of Minnesota Press, Minneapolis, 1997, pp. 121-146.
[16] D. Bigo, 'When Two Become One: Internal and External Securitisations in Europe,' *International Relations Theory and the Politics of European Integration: Power, Security and Community*, M. Kelstrup & M.C. Williams (eds), Routledge, London, 2000, pp. 171-204; D. Bigo, 'Security and Immigration: Towards a Critique of the Governmentality of Unease,' *Alternatives*, Vol. 27, 2002, pp. 63-92; D. Bigo, 'Globalized (in)Security: The Field and the Ban-opticon,' *Illiberal Practices of Liberal Regimes—The (in)Security Games*, D. Bigo & A. Tsoukala (eds), l'Harmattan, Paris, 2006, pp. 5-49.
[17] Obviously, these are just *models* and the incorporation of all variables contradicts the pursuit of parsimony; see K. Waltz, *Theory of International Politics*, Addison-Wesley, Reading/Massachusetts, 1979. Nevertheless, the question has to be asked as to how useful models are, which mostly exclude central topics (the human individual), and if it wouldn't be better to include these topics to a greater degree.
[18] A. Giddens, *The Constitution of Society: Outline of the Theory of Structuration*, University of California Press, Berkeley/Los Angeles, 1984.
[19] R.L. Doty, 'Aporia: A Critical Exploration of the Agent-Structure Problematique in International Relations Theory,' *European Journal of International Relations*, Vol. 3, September 1997, pp. 365-392; also see Bourdieu's Habitus Concept: P. Bourdieu, *Outline of a Theory of Practice*, Cambridge University Press, Cambridge, 1977; P. Bourdieu, *Distinction: A Social Critique of the Judgment of Taste,* Harvard University Press, Cambridge/Massachusetts/London, 1984.
[20] United Nations Development Programme, *Human Development Report 1994*, p. 22.
[21] Ibid., pp. 24-33.
[22] J. Leaning, in Burgess & Owen, op. cit., p. 354.
[23] S. Van Evera, *Guide to Methods for Students of Political Science*, Cornell University Press, Ithaca and London, 1997, p. 20; A. Mack, in Burgess & Owen, op. cit., p. 367.
[24] Paris, op. cit., p. 92.
[25] H. Müller, *'Das Leben selbst ist lebensgefährlich', Kritische Anmerkungen zum 'erweiterten Sicherheitsbegriff,'* HSFK-Standpunkte 4/1997, Hessische Stiftung Friedens- und Konfliktforschung, Frankfurt a.M., 1997, Author's translation of the article's title.
[26] K. Krause, in Burgess & Owen, op. cit., p. 368; see also Ayoob, op. cit., p. 121: 'the often indiscriminate broadening of the definition of security

threatens to make the concept so elastic as to render it useless as an analytical tool.'

[27] For an overview see O. Wæver, 'Aberystwyth, Paris, Copenhagen: New 'Schools' in Security Theory and their Origins between Core and Periphery,' Paper presented at the annual meeting of the International Studies Association, Montreal, March 17-20, 2004, Viewed on 8 December 2007, <http://www.allacademic.com/meta/p74461_index.html>.

[28] O. Wæver, 'Securitization and Desecuritization,' *On Security*, R. Lipschutz (ed), Columbia University Press, New York, 1995, pp. 46-86.

[29] Ibid., p. 47.

[30] Ibid., p. 48.

[31] Ibid., p. 50.

[32] T. Körner, '... Rumlungern, Alkohol und... und... und..., Sicherheits- und Ausschließungsstrategien in der Münchner Innenstadt,' *Öffentliche Räume—öffentliche Träume. Zur Kontroverse über die Stadt und die Gesellschaft*, C.G. Wiegandt (ed), Lit Verlag, Berlin, 2006, pp. 47-59.

[33] Wæver, op. cit., p. 54.

[34] Ibid., p. 63.

[35] Ibid., p. 57.

[36] United Nations Development Programme, *Human Development Report 1994*, pp. 30-31.

[37] I owe the emphasis on this specification to Bob Brecher and thank him for this helpful comment.

[38] A. Mack (ed), *Human Security Report 2005: War and Peace in the 21st Century*, Published for the Human Security Centre, University of British Columbia, Oxford University Press, New York, 2005, p. 257.

[39] Wæver, op. cit., pp. 50-51; R Paris, in Burgess & Owen, op. cit., p. 371.

[40] Of course, conceptualising security as an attributive-only variable would theoretically be possible, too—that is because no positive, real (in a scientific realist sense), empirical findings are concerned, but concepts; so there is no truth in this case. But there is, however, the question of better conceptual functionality—or in Glasersfeld's words: it's about the 'viability,' and this speaks in favour of an exclusive description and operationalisation of security; see E. Glasersfeld, 'Konstruktion der Wirklichkeit und des Begriffs der Objektivität,' *Einführung in den Konstruktivismus*, H. Foerster, E. Glasersfeld, P.M. Hejl, S.J. Schmidt & P. Watzlawick (eds), Piper, Munich, 1992, pp. 9-39.

[41] Inasmuch it seems questionable, whether even the vague term of 'political leitmotif,' which Werthes and Bosold use, is overrated; see S. Werthes & D. Bosold, *Das 'Human Security'-Paradigma und seine Dilemmata*, Vortrag auf der AFK-Nachwuchstagung 'Krieg, Gewalt und der prekäre Frieden,'

Europa-Institut Bocholt, 14-16 January 2005, Viewed on 18 December 2007, <http://www.staff.uni-marburg.de/~bosold/pdf/Werthes-Bosold_AFK.pdf>; S. Werthes & D. Bosold, 'Caught between Pretension and Substantiveness—Ambiguities of Human Security as a Political Leitmotif,' *Human Security on Foreign Policy Agendas: Changes, Concepts and Cases*, T. Debiel & S. Werthes (eds), INEF Report 80/2006, Institute for Development and Peace, University of Duisburg-Essen, Duisburg, 2006, Viewed on 23 November 2007, <http://inef.uni-due.de/page/documents/Report80.pdf>, pp. 21-38; M.C. Williams, 'Modernity, Identity and Security: a Comment on the Copenhagen Controversy,' *Review of International Studies*, Vol. 24, July 1998, pp. 435-439.

[42] R. Floyd, 'Human Security and the Copenhagen School's Securitization Approach: Conceptualizing Human Security as a Securitizing Move,' *Human Security Journal/Revue de la Sécurité Humaine*, Issue 5, Winter 2007, pp. 38-49.

[43] For a complex of topics similar to humanitarian intervention, see Avery Plaw's contribution in this book. He discusses the question of the legality of targeted killing, which—although on another level and from a different point of view—is related with the issue of just war.

[44] Whereas, as Walzer criticises, it has to come about eventually and must not be postponed ad infinitum; M. Walzer, 'The Argument about Humanitarian Intervention,' *Dissent*, Vol. 49, Winter 2002, p. 33.

[45] Ibid., p. 31.

[46] Ibid., p. 30; furthermore, Walzer writes about the limits of the reasons for humanitarian interventions: 'The new regime doesn't have to be democratic or liberal or pluralist or (even) capitalist. It doesn't have to be anything, except non-murderous. When intervention is understood in this minimalist fashion, it may be a little easier to see it through.' (Ibid., p. 36).

[47] United Nations Development Programme, *Human Development Report 1994*, p. 22.

[48] International Committee of the Red Cross (ICRC) (ed), *Iraq: No Let-up in the Humanitarian Crisis*, ICRC, Geneva, 2008, Viewed on 24 November 2011, <http://www.icrc.org/eng/assets/files/other/icrc-iraq-report-0308-eng.pdf>.

[49] C. Schmitt, *The Concept of the Political*, Rutgers University Press, New Brunswick, 1976.

[50] R. Taureck, *Securitisation Theory—The Story so Far: Theoretical Inheritance and What it Means to be a Post-Structural Realist*, Paper for presentation at the 4th annual CEEISA Convention, University of Tartu, 25-27 June 2006, Viewed on 13 December 2007, rev. 2007-12-13,

<http://www.ceeisaconf.ut.ee/orb.aw/class=file/action=preview/id=164452/TAURECK.doc>.

[51] e.g., C. Aradau, *Desecuritize and Despair! The Copenhagen School Revisited*, CEU Working Paper, Central European University, Budapest, 2003; J. Huysmans, 'Migrants as a Security Problem: Dangers of 'Securitizing' Societal Issues,' *Migration and European Integration: The Dynamics of Inclusion and Exclusion*, R. Miles & D. Thranhardt (eds), Pinter, London, 1995, pp. 53-72; Wæver, 'Securitization and Desecuritization,' op. cit.; O. Wæver, 'The EU as a Security Actor: Reflections from a Pessimistic Constructivist on Post-Sovereign Security Orders,' *International Relations Theory and the Politics of European Integration: Power, Security and Community*, M. Kelstrup & M.C. Williams (eds), Routledge, London, 2000, pp. 250-294.

[52] Huysmans terms these strategies *objectivist*, *constructivist* and *deconstructivist*, and prefers the last one of them; see Huysmans, op. cit., pp. 65-67; Roe, op. cit., pp. 285-286.

[53] P. Berger & T. Luckmann, *The Social Construction of Reality: A Treatise in the Sociology of Knowledge*, Anchor Books, Garden City/New York, 1966, p. 59.

[54] Roe, op. cit., p. 287.

Bibliography

Agamben, G., *State of Exception*. University of Chicago Press, Chicago, 2005.

Alkire, S., *A Conceptual Framework for Human Security*. CRISE Working Paper 2, Centre for Research on Inequality, Human Security and Ethnicity (CRISE), University of Oxford, Oxford, 2003.

Aradau, C., *Desecuritize and Despair! The Copenhagen School Revisited*. CEU Working Paper, Central European University, Budapest, 2003.

Axworthy, L., *Human Security: Safety for People in a Changing World*. Concept Paper prepared for the Canadian Department of Foreign Affairs and International Trade, April 1999, Viewed on 13 December 2007, <http://www.summit-americas.org/Canada/HumanSecurity-english.htm>.

Bajpai, K., *Human Security: Concept and Measurement.* Kroc Institute Occasional Paper #19:OP1, The Joan B. Kroc Institute for International Peace Studies, University of Notre Dame, Notre Dame, August 2000, Viewed Viewed on 24 November 2011, <http://www.hegoa.ehu.es/dossierra/seguridad/Human_security_concept_and_measurement.pdf>.

Baldwin, D.A., 'The Concept of Security.' *Review of International Studies.* Vol. 23, January 1997, pp. 5-26.

Beck, U., *Risk Society: Towards a New Modernity.* Sage, New Delhi, 1992.

—, *World Risk Society.* Polity Press, Cambridge, 1999.

—, *Weltrisikogesellschaft. Auf der Suche nach der verlorenen Sicherheit.* Suhrkamp, Frankfurt am Main, 2007.

Berger P. & Luckmann, T., *The Social Construction of Reality: A Treatise in the Sociology of Knowledge.* Anchor Books, Garden City/New York, 1966.

Bigo, D., 'When Two Become One: Internal and External Securitisations in Europe.' *International Relations Theory and the Politics of European Integration: Power, Security and Community.* Kelstrup, M.& Williams, M.C. (eds), Routledge, London, 2000.

—, 'Security and Immigration: Towards a Critique of the Governmentality of Unease.' *Alternatives.* Vol. 27, 2002, pp. 63-92.

—, 'Globalized (in)Security: The Field and the Ban-opticon.' *Illiberal Practices of Liberal Regimes: The (in)Security Games.* Bigo, D. & Tsoukala, A. (eds), L'Harmattan, Paris, 2006.

Bourdieu, P., *Outline of a Theory of Practice.* Cambridge University Press, Cambridge, 1977.

—, *Distinction: A Social Critique of the Judgment of Taste.* Harvard University Press, Cambridge/London, 1984.

Brandão, A.P., *The Redefinition of Security: An Assessment.* Paper to be presented at The Third Pan-European International Relations Conference and Joint Meeting with the International Studies Association, Vienna, 16-19 September 1998.

Burgess, P.J. & Owen, T. (eds), 'Special Section: What is Human Security?' *Security Dialogue*. Vol. 35, September 2004, pp. 345-387.

Buzan, B., *People, States and Fear: An Agenda for International Security Studies in the Post-Cold War Era*. Harvester Wheatsheaf, New York, 1991.

Buzan, B. & Wæver, O., 'Slippery? Contradictory? Sociologically Untenable? The Copenhagen School Replies.' *Review of International Studies*. Vol. 23, April 1997, pp. 241-250.

—, *Regions and Powers: The Structure of International Security*. Cambridge University Press, Cambridge, 2003.

Buzan, B., Wæver, O. & DeWilde, J., *Security: A New Framework for Analysis*. Lynne Rienner, Boulder, 1998.

Commission on Human Security, *Human Security Now*. Commission on Human Security, New York, 2003, Viewed on 24 November 2011, <http://www.policyinnovations.org/ideas/policy_library/data/01077/_res/id=sa_File1/>.

Croft, S. & Terriff, T. (eds), *Critical Reflections on Security and Change*. Frank Cass, London/Portland, 2000.

Daase, C., 'Internationale Risikopolitik. Ein Forschungsprogramm für den sicherheitspolitischen Paradigmenwechsel.' *Internationale Risikopolitik. Der Umgang mit neuen Gefahren in den internationalen Beziehungen*. Daase, C., Feske, S. & Peters, I. (eds), Nomos, Baden-Baden, 2002.

Debiel, T. & Werthes, S. (eds), *Human Security on Foreign Policy Agendas: Changes, Concepts and Cases*. INEF Report 80/2006, Institute for Development and Peace, University of Duisburg-Essen, Duisburg, 2006, Viewed on 23 November 2011, <http://inef.uni-due.de/page/documents/Report 80.pdf>.

Derrida, J., 'The Deconstruction of Actuality' (interview). *Radical Philosophy*. Iss. 68, Autumn 1994, pp. 28-41.

Dorn, W., *Human Security: An Overview*. Pearson Peacekeeping Centre, Ontario, 2001, Viewed 24 November 2011, <http://walterdorn.org/pub/23>.

Doty, R.L., 'Aporia: A Critical Exploration of the Agent-Structure Problematique in International Relations Theory.' *European Journal of International Relations*. Vol. 3, September 1997, pp. 365-392.

Floyd, R., 'Human Security and the Copenhagen School's Securitization Approach: Conceptualizing Human Security as a Securitizing Move.' *Human Security Journal/Revue de la Sécurité Humaine*. Iss. 5, Winter 2007, pp. 38-49.

Galtung, J., 'Violence, Peace, and Peace Research.' *Journal of Peace Research*. Vol. 6, 3/1969, pp. 167-191.

Giddens, A., *The Constitution of Society: Outline of the Theory of Structuration*. University of California Press, Berkeley/Los Angeles, 1984.

Grayson, K., 'Securitization and the Boomerang Debate: A Rejoinder to Liotta and Smith-Windsor.' *Security Dialogue*. Vol. 34, September 2003, pp. 337-343.

Guzzini, S. & Jung, D. (eds), *Contemporary Security Analysis and Copenhagen Peace Research*. Routledge, London/New York, 2004.

Huysmans, J., 'Migrants as a Security Problem: Dangers of 'Securitizing' Societal Issues.' *Migration and European Integration: The Dynamics of Inclusion and Exclusion*. Pinter, London, 1995.

International Committee of the Red Cross (ICRC), *Iraq: No Let-up in the Humanitarian Crisis*. ICRC, Geneva, 2008, Viewed on 24 November 2011, <http://www.icrc.org/eng/assets/files/other/icrc-iraq-report-0308-eng.pdf>.

King, G. & Murray, C.J.L., 'Rethinking Human Security.' *Political Science Quarterly*. Vol. 116, Winter 2001-2002, pp. 585-610.

Körner, T., 'Rumlungern, Alkohol und... und... und....' Sicherheits- und Ausschließungsstrategien in der Münchner Innenstadt.' *Öffentliche Räume—öffentliche Träume. Zur Kontroverse über die Stadt und die Gesellschaft*. Wiegandt, C.G. (ed), Lit Verlag, Berlin, 2006, pp. 47-59.

Krause, K. & Williams, M.C. (eds), *Critical Security Studies*. University of Minnesota Press, Minneapolis, 1997.

Lipschutz, R.D. (ed), *On Security*. Columbia University Press, New York, 1995.

Mack, A. (ed), *Human Security Report 2005: War and Peace in the 21st Century*. Published for the Human Security Centre, University of British Columbia, Oxford University Press, New York, 2005.

McSweeney, B., 'Identity and Security: Buzan and the Copenhagen School.' *Review of International Studies*. Vol. 22, January 1996, pp. 81-93.

—, 'Durkheim and the Copenhagen School: A Response to Buzan and Wæver.' *Review of International Studies*. Vol. 24, January 1998, pp. 137-140.

Millar, K., 'A Human Security Analysis of the War in Iraq.' *Human Security Journal/Revue de la Sécurité Humaine*. Iss. 2, June 2006, pp. 47-63.

Müller, H., *'Das Leben selbst ist lebensgefährlich'. Kritische Anmerkungen zum 'erweiterten Sicherheitsbegriff'*. HSFK-Standpunkte 4/1997, Hessische Stiftung Friedens- und Konfliktforschung, Frankfurt a.M., 1997.

Newman, E., 'Human Security and Constructivism.' *International Studies Perspectives*. Vol. 2, August 2001, pp. 239-251.

Paris, R., 'Human Security: Paradigm Shift or Hot Air?' *International Security*. Vol. 26, Fall 2001, pp. 87-102.

Roberts, D., 'Human Security or Human Insecurity? Moving the Debate Forward.' *Security Dialogue*. Vol. 37, June 2006, pp. 249-261.

Roe, P., 'Securitization and Minority Rights: Conditions of Desecuritization.' *Security Dialogue*. Vol. 35, September 2004, pp. 279-294.

Roosevelt, F.D., *The Annual Message to the Congress*. 6 January 1941, Viewed on 24 November 2011, <http://www.presidency.ucsb.edu/ws/index.php?pid=16092#axzz1eeXELadb>.

Rothschild, E., 'What is Security?' *Daedalus*. Vol. 124, Summer 1995, pp. 53-98.

Sartori, G., 'Concept Misinformation in Comparative Politics.' *American Political Science Review*. Vol. 64, December 1970, pp. 1033-1053.

Schmitt, C., *The Concept of the Political*. Rutgers University Press, New Brunswick, 1976.

Stritzel, H., 'Towards a Theory of Securitization: Copenhagen and Beyond.' *European Journal of International Relations*. Vol. 13, September 2007, pp. 357-383.

Tadjbakhsh, S., *Human Security: Concepts and Implications with an Application to Post-Intervention Challenges in Afghanistan*. Les Etudes du CERI—n° 117-118, Centre for Peace and Conflict Resolution, Sciences Po, Paris, 2005, Viewed on 24 November 2011, <http://www.glow-boell.de/media/de/txt_rubrik_3/Ulbert_autorisiert.pdf>.

Taureck, R., *Securitisation Theory—The Story so Far: Theoretical Inheritance and What it Means to be a Post-Structural Realist*. Paper for presentation at the 4th annual CEEISA Convention, University of Tartu, 25-27 June 2006, Viewed on 13 December 2007, Rev. 2007-12-13, <http://www.ceeisaconf.ut.ee/orb.aw/class=file/action=preview/id=164452/TAURECK.doc>.

Ulbert, C., *Human Security—ein brauchbares Konzept für eine geschlechtergerechte außen- und sicherheitspolitische Strategie?* Vortrag auf der Fachtagung des Feministischen Instituts der Heinrich-Böll-Stiftung in Kooperation mit der Friedrich-Ebert-Stiftung und dem Frauensicherheitsrat, Abgeordnetenhaus Berlin, Berlin 24/25, October 2003, Viewed on 17 December 2007, <http://www.atasp.de/downloads/ulbert_hs.pdf>.

United Nations Development Programme (UNDP), *Human Development Report 1994*. UNDP, New York, 1994, Viewed on 24 November 2011, <http://hdr.undp.org/en/reports/global/hdr1994/chapters/>.

—, *Human Development Report 2005*. UNDP, New York, 2005, Viewed on 24 November 2011, <http://hdr.undp.org /reports/global/hdr2005/chapters/>.

Van Evera, S., *Guide to Methods for Students of Political Science*. Cornell University Press, Ithaca/London, 1997.

Waltz, K.N., *Theory of International Politics*. Addison-Wesley, Reading, Massachusetts, 1979.

Walzer, M., *Just and Unjust Wars: A Moral Argument with Historical Illustrations*. Basic Books, New York, 1977.

—, 'The Argument about Humanitarian Intervention.' *Dissent*. Vol. 49, Winter 2002, pp. 29-37.

Wæver, O., 'Securitization and Desecuritization.' *On Security*. Lipschutz, R. (ed), Columbia University Press, New York, 1995.

—, 'The EU as a Security Actor: Reflections from a Pessimistic Constructivist on Post-Sovereign Security Orders.' *International Relations Theory and the Politics of European Integration: Power, Security and Community*. Kelstrup, M. & Williams, M.C. (eds), Routledge, London, 2000, pp. 250-294.

—, 'Aberystwyth, Paris, Copenhagen: New 'Schools' in Security Theory and their Origins between Core and Periphery.' Paper presented at the annual meeting of the International Studies Association, Montreal, March 17-20, 2004, Viewed on 8 December 2007, <http://www.allacademic.com/meta/p74461_index.html>.

Werthes, S. & Bosold, D., *Das 'Human Security'-Paradigma und seine Dilemmata*. Vortrag auf der AFK-Nachwuchstagung 'Krieg, Gewalt und der prekäre Frieden,' Europa-Institut Bocholt, 14-16 January 2005, Viewed on 18 December 2007, <http://www.staff.uni-marburg.de/~bosold/pdf/Werthes-Bosold_AFK.pdf>.

—, 'Caught between Pretension and Substantiveness: Ambiguities of Human Security as a Political Leitmotif.' *Human Security on Foreign Policy Agendas: Changes, Concepts and Cases*. Debiel, T. & Werthes, S. (eds), INEF Report 80/2006, Institute for Development and Peace, University of Duisburg-Essen, Duisburg, 2006, Viewed on 23 November, <http://inef.uni-due.de/page/documents/Report80.pdf>, pp. 21-38.

Williams, M.C., 'Modernity, Identity and Security: A Comment on the Copenhagen Controversy.' *Review of International Studies*. Vol. 24, July 1998, pp. 435-439.

Wolfers, A., 'National Security as an Ambiguous Symbol.' *Political Science Quarterly*. Vol. 67, December 1952, pp. 481-502.

Benjamin Rampp is a research associate at the Chair of General Sociology, University of Trier, Germany and at the International Centre for Ethics in the Sciences and Humanities (IZEW) (Section Ethics and Culture—Security Ethics), University of Tübingen, Germany.

Towards an Enhanced Understanding of Diplomacy as the Business of Peace

Stuart Murray

Abstract
In this chapter it is proposed that the traditional view of diplomacy is an archaic vision of the 'engine room of international relations.'[1] This rhetoric, it is argued, is parochial and does not match the realities of the modern, twenty-first century diplomatic environment where pluralistic, pacific and polylateral networks of diplomacy are thriving. In the modern diplomatic environment, the activity of diplomacy should be viewed as the business of multi-actor peace and not only the handmaiden of the occasionally belligerent state. New, radical and innovative views on diplomacy are flooding the diplomacy studies field. This chapter will summarise these views into three distinct Schools of diplomatic thought—Traditional, Nascent and Innovative—which allude to this positive growth in scholarship. This chapter will argue—centrally—that an enhanced understanding of diplomacy is the best chance society has to replace the norm of conflict with that of stability and peace achieved through diplomatic functions.

Key Words: Diplomacy, Diplomacy Studies, peace, epistemology, taxonomy, theory.

*****

> Negotiators have often worked by means of promises, appeals to interest, attempts at striking a bargain, devices of cajolery. They have resorted sometimes to taunts and to bullying, sometimes to quiet blackmail or impudent bluff. Even the threat of war may be one of the counters which the diplomat uses and this itself might be merely a piece of bluff or might call for some delicate interpretation. [...] Diplomacy may include anything short of actual war, therefore and sometimes the kindest thing that one can say of it is that it is better than having the guns actually firing. - Sir Herbert Butterfield, *Diplomacy*

By their very nature, diplomacy and war are means to the ends of statecraft as well as channels by which governments press their agendas onto others. Diplomacy is often thought of as the peaceful alternative to violence but in fact diplomacy serves to prepare as often as to avoid war. It is an important part of waging war, often makes the difference in who wins and nearly always codifies wars' results. This familiar rhetoric constitutes the traditional view of diplomacy.[2]

In this chapter it is proposed that this traditional view is an archaic vision of the 'engine room of international relations.'[3] This rhetoric, it is argued, is parochial and does not match the realities of the modern, twenty-first century diplomatic environment where pluralistic, pacific and polylateral networks of diplomacy are thriving. In the modern diplomatic environment, the activity of diplomacy should be viewed as the business of multi-actor peace and not only the handmaiden of the occasionally belligerent state.

This chapter will demonstrate that the diplomatic studies field is distancing itself from the centuries-old traditional views of diplomacy (that is, diplomacy as an exclusive state tool or war by other means). New, radical and innovative views on diplomacy are flooding the diplomacy studies field. This chapter will summarise these views into three distinct Schools of diplomatic thought—Traditional, Nascent and Innovative—which allude to this positive growth in scholarship.

This chapter will argue—centrally—that an enhanced understanding of diplomacy is the best chance society has to replace the norm of conflict with that of stability and peace achieved through diplomatic functions. An enhanced understanding of the theory, practice and application of diplomacy—as this chapter attempts—can only lead to a more diplomatic and therefore more peaceful international relations system.

1. Why Study Diplomacy?

Diplomacy is an increasingly vital activity in the modern international relations system. It has always been an important component of the system from the Greeks to the Byzantines to the Italian city-states model.[4] However, since Westphalia diplomacy's significance to international stability has been underestimated. It has been labelled as a rather dour, sacrosanct and hermetically sealed realm best left to the professionals: the diplomats. For many centuries, diplomacy has also been plagued by an ill-founded and over simplistic stereotype where 'among general publics, a well-developed image exists of [a] privileged elite pursuing exciting and prestigious careers without paying parking tickets and with varying degrees of effectiveness.'[5] The only time there appeared genuine interest in diplomacy was when it failed.

This stereotypical and shallow view of diplomacy is frustrating as it suggests diplomacy is a rather quaint, archaic and inept activity. Yet such views are far removed from the realities of the twenty-first century diplomatic environment.

In the physical sense, diplomacy remains the 'engine room of international relations' in the complex 21st century.[6] Martin Wight accurately describes diplomacy as the 'master-institution of international relations' and even Sir Robert Peel's observation that diplomacy is 'the great engine used by civil society for the purposes of maintaining peace' bears modern relevance.[7]

In the twenty-first century, the study and practice of diplomacy are enjoying a renaissance. Many international problems are solved through communication, accurate information gathering and dissemination, dialogue, compromise and negotiation: the pillars of diplomacy.[8] Today, state and non-state actors appear more willing to engage in diplomacy than all out hostility. To promote this preferable global condition it is important to enhance understanding of diplomacy, both inside and outside of the diplomacy studies field.

2. What is Diplomacy?

To promote an enhanced understanding of diplomacy it is important to begin with 'what we know' of diplomacy. State-qua-state, traditional diplomacy is the most readily identifiable type of diplomacy and is an appropriate departure point. Each of the following traditional and popular definitions impress upon the reader a similar message: that the state is central to any theoretical understanding of diplomacy and that diplomacy, largely, concerns the business of peace.

Hedley Bull defines diplomacy as 'the conduct of relations between sovereign states with standing in world politics by official agents and by peaceful means.'[9] G.R. Berridge also packages diplomacy simply and in statist terms. He writes that 'diplomacy is the term given to the official channels of communication employed by the members of a system of states.'[10] Berridge recognises that 'diplomacy is the most important institution of our society of states' before adding that 'the chief purpose of diplomacy is to enable states to secure the objectives of their foreign policies without resort to force, propaganda, or law.'[11]

Sir Ernest Satow develops the peaceful aspect of diplomacy further. He defines diplomacy as 'the application of intelligence or tact to the conduct of relations between the governments of independent states; or more briefly still, the conduct of business between states by *peaceful* means.'[12] Satow adds that 'persuasive argument, if applied skilfully and sensitively at the right time, may achieve a better result than persuasion too obviously backed by the threat of force.'[13] He claims that diplomacy is 'the best means devised by civilisation for preventing international relations from being governed by force alone.'[14] In other words, diplomacy is concerned with minimising friction between states and mitigating the anarchical nature of the IR system, which can occasionally erupt in conflict. For Satow, diplomacy and conflict are mutually exclusive, where a rational state actor would view diplomacy as a more pacific and preferable alternative to conflict. Therefore, diplomacy and war can be interpreted as exclusive forces, rather than 'activities which can take place simultaneously or may follow from one another.'[15]

This traditional image of diplomacy has several synonyms: 'statist,' 'state-centric,' or 'rationalist' being common.[16] This chapter prefers the label 'Traditionalism.' The tradition in this case is to continue to emphasize the centrality of the state to diplomacy. Continuity then allows each generation of Traditionalists to build on the foundations laid by their theoretical forefathers. Each of the Traditionalists relies on, develops and expresses an admiration for the work of their predecessors.[17]

Traditionalists share five common assumptions that allude to a distinct way of thinking and writing on diplomacy: together these comprise a Traditional School of thought. First, Traditionalists describe diplomacy as an exclusive State function. They infer 'that diplomacy is the privileged domain of professional diplomats, conducted almost exclusively by Foreign Service personnel and officials from Foreign Ministries.'[18] Second, Traditionalists interpret diplomacy as the study of the international realm of sovereign states, with the central purpose of diplomacy being to overcome the anarchical nature of that system and to facilitate peaceful relationships amongst sovereign states through familiar, historical channels of diplomacy. Traditionalists, thirdly, concentrate on diplomacy's role in relation to a classic political-military agenda, which includes issues such as force balances, demarcation of territories, arms control negotiations and alliance cohesion. Fourth, Traditionalists consider diplomatic and political history as central to their School of thought. The study of diplomacy, they argue, demands an embracing of the distant as well as the recent past. Fifth, Traditionalists write prescriptive guides to diplomacy where they theorise on the practice of diplomacy. Satow's *Guide to Diplomacy* (1957), Berridge's *Diplomacy: Theory and Practice* (2002) and Rana's *The 21st Century Ambassador* (2004) can be classified as such. These books are 'manuals of diplomatic procedure, in the tradition of the guides that made their appearance early in the European system and continue to dominate the field.'[19] They convey a view of diplomacy as a specialised skill, the vocation of the select few and are beneficial for the novice entering the profession. Traditionalism has dominated the diplomacy studies field for centuries. However, this approach to diplomacy is not without fault. Indeed, Traditionalism's inherent exclusivity (of actor) and parochialism suggests the deficiencies to this School.

One central deficiency concerns the works that dominate the canon of diplomacy studies which are prescriptive, state-centric and, largely, archaic. The diplomacy studies field's 'classics' are Wicquefort, De Callieres and Satow, for example, and these works are prominent in any environment where diplomacy is studied.[20] The canon of diplomacy studies is dominated by these guides on 'how to' conduct diplomacy at the official, state level.

Such 'conventional accounts' of diplomacy usually include:

> a narration of the progressive story of diplomatic history; the organisation of the Ministry of Foreign Affairs; the formulation of foreign policy; the functions of the embassy; the qualities of the diplomat; and different accounts dealing with issues ranging from negotiation to immunities and from international trade or law to etiquette and protocol.[21]

While these prescriptive guides offer novices an insight into the practicalities and rules of official diplomatic interaction, they add little to a canon reflective of modern diplomacy. As a result, murmurs of discontent can at last be heard. The preference to write on diplomacy in a prescriptive fashion led Watson to lament that 'no good book on the wider aspects of diplomacy, as distinct from foreign policies of individual states and the details of diplomatic practice, has been written since Harold Nicolson's classic *Diplomacy* (1939).'[22] More recently, Lee and Hudson complain that:

> The conceptual framework of diplomacy, the definition of diplomacy and the systemic environment of diplomacy has been constant in diplomacy studies from Wicquefort (1606-1682). In essence, the very idea of diplomacy—that it is a dialogue between states in an anarchic systemic structure of independent political units—has not changed all that much during some three hundred and fifty years of scholarship.[23]

That the diplomacy studies field's theoretical understanding of diplomacy has hardly altered during three and half centuries is enlightening.[24] The ongoing reliance upon the classic texts could suggest that there is an absence of original and innovative scholarship on modern diplomacy.

Fortunately, the traditional approach to writing and thinking on diplomacy is being supplemented. The increasing complexity of the modern diplomatic environment is creating a demand for more innovative works on diplomacy, forcing a foundational alteration of the canon. Since the end of the Cold War a plethora of innovative works on diplomacy have emerged, all of which challenge a central traditional tenet: that the state holds a monopoly over diplomacy.

3. Towards an Enhanced Understanding of Diplomacy as the Business of Peace

The renaissance in diplomacy studies can be confirmed by recognising the explosion in new, radical and innovative views on modern diplomacy. Central to this evolution is the promise that rewarding insight awaits if the field disassociates—but does not abandon—the axiomatic notion that diplomacy must be linked to its state master. Such emerging views on what constitutes modern diplomacy can be classified into two further Schools of diplomatic thought.

A. The Nascent School

The second school of diplomatic thought, the Nascent School, emerged to challenge the dominant Traditionalist School. Scholars from this group focus on emerging forms of alternate diplomacy: hence the label Nascent. This type of diplomacy has several synonyms: new, unofficial, unconventional, or track two diplomacy, for example. The premise that the 'the diplomatic expertise for dealing successfully with conflict and peace-making does not reside solely within government personnel or procedures' is central to this School.[25]

Nascent scholars, like Traditionalists, share common assumptions and generalisations, allowing us to speak of a distinct School. There are five in total. First, Nascent theorists challenge the notion that diplomacy should be interpreted in a rigid, precise or authoritative fashion, concentrating on the role of the state. Nascent scholars, second, view the state and its diplomacy as blocking change to a more pacific international relations system. Nascent writers can be described as 'those who regard the state as an obstacle to world order;' to them, 'the development of an alternative diplomacy, embracing non-governmental organisations (NGOs) and transnational movements, offers the prospect of an international order transcending the state system.'[26] Third, Nascent scholars argue that the traditional diplomatic institution is in a period of crisis.[27] Nascent authors believe that traditional diplomacy is ill-suited to twenty-first century problems and that state-centric diplomacy, both as a vocation and an area of theoretical focus, is defunct. The crisis of the traditional diplomatic institution leads Nascent scholars to suggest, fourth, that perhaps it is obsolete. The same criticism is levelled at the state-centric diplomatic writing that dominates the canon. Nascent scholars argue that this parochial opinion fails to account for the multi-actor complexity of the modern diplomatic environment. The obsolescence attributed to the traditional way of writing and thinking on diplomacy has led Nascent scholars, fifth, to focus theoretical efforts on alternate diplomatic actors. To understand modern diplomacy, claim Nascent scholars, means first to acknowledge the multi-actor nature of the modern diplomatic environment and, second, to theorise on the diplomacy that unconventional actors practise within that environment.

From these five common assumptions a definition of Nascent diplomacy can be introduced.[28] Nascent diplomacy can be framed as a reaction to states' tardiness in embracing an unconventional or lower agenda that has grown in importance since the end of the Cold War. Nascent

Diplomacy is characterised by the fostering of equitable, networked and stable relationships amongst non-state actors. These actors share interdependent, low political goals and diplomatically exchange resources, expertise and knowledge (information) in pursuit of unconventional goals such as human rights or aid and development issues. Nascent or non-state diplomatic actors have capitalised on state deficiencies to promote their agenda, their position in the international relations hierarchy and their alternate but effective political and diplomatic techniques for tackling global problems, which states have been sluggish in addressing.

This School, like Traditionalism, in not without its faults. Perhaps the most obvious falsifying instance concerns its dismissive attitude toward the incumbent state and its traditional diplomatic institution; simply, if these entities are obsolete, in crisis or irrelevant then why do they continue to exist? To deny Traditionalism or state-centrism is to deny an actuality of the modern diplomatic environment: the omnipresence of traditional diplomacy.

This School, however, is central to developing a deeper understanding of modern diplomacy in that it encourages scholars to develop further questions, hypotheses, which, in turn, lead to novel facts concerning modern diplomacy. By default, the Nascent School pushes the field towards 'novel' areas of diplomacy and endorses the notion that the surface of diplomacy has only just been scratched.

B. The Innovative School

The third School, the Innovative School, emerged (largely) as a result of the polemic relationship between the Traditional and Nascent Schools. Where Traditionalists and Nascent scholars are concerned, such approaches still yield cantilevered bridges since their builders do not significantly relax the fundamental assumptions that distinguish the contending research traditions.[29] Therefore, a middle ground, privileging both the state and the non-state, is conspicuous by its absence. The innovative School emerged to occupy the middle ground.

Scholars from this group also share four common assumptions or generalisations which allude to a third distinct School.[30] First, the Innovators are critical of the divisionary relationship between the Traditional and Nascent Schools. Often, within diplomacy studies, state and non-state actors are portrayed as:

> inhabiting different environments, working to different rulebooks and occupying very different positions on the scale of importance in world politics. They exist, therefore, in two solitudes with little or no interaction between their worlds.[31]

The exclusive views of Traditionalists and Nascent scholars are damaging, as they encourage competition of opinion at the expense of accuracy. Such competition, according to the Innovators, means that an impartial body of diplomatic theory truly reflective of the modern diplomatic environment has yet to emerge.

Highlighting the futility of this condition is a second common assumption. For the Innovators, polarisation of diplomatic thought forces the scholar to make an either/or choice. For example, Hocking warns that the study and theory of diplomacy 'has become an icon for the advocates of two competing perspectives on international politics, the state-centric and world society views.'[32] This 'confused and unfocussed' dialogue has resulted in 'two lines of divergent argument' (state versus non-state) which are in danger of bogging the diplomacy studies field in 'sterile and unproductive debate.'[33] One gets the impression that the Innovators think the time of Traditional and Nascent diplomatic scholars could be better spent.

Banishing or dismissing such either/or, state versus non-state rhetoric is a third assumption of the Innovators. They do so through a continuous and objective (re)appraisal of the state/non-state relationship, which, they argue, is distinctly non-adversarial, symbiotic and complementary.[34] The propensity to moderate and incorporate balance into their way of writing and thinking on diplomacy is the hallmark of this group of scholars.

Of the Innovators, Melissen notes that they 'try to provide insights into a transforming diplomatic landscape and the changing modalities and forms of diplomacy within.'[35] They aim to incorporate observations on non-state diplomatic actors into their way of thinking about diplomacy. Within the Innovative School, both the state and the non-state are given equal weighting. This practical balance is a fourth common assumption shared by the Innovators.

In this context, there does not appear to be anything particularly 'innovative' to the Innovative Diplomacy School. After all, knowledge on the diplomacy of state and non-state actors is readily available in the diplomacy studies literature. However, it is not the knowledge that is innovative. Rather, an eclectic approach which privileges both the state and the non-state is novel in diplomacy studies. This orientation suggests a 'strong demonstration of renewed theoretical innovation in the field, innovation that avoids old patterns and old labels.'[36]

An Innovative definition of diplomacy confirms this desire to avoid familiar theoretical terrain and to think differently on modern diplomacy. Innovators infer that the term diplomacy is applicable to groups, not only states but also nascent actors such as NGOs or multi-national corporations (MNCs) who have an equal interest in a stable and peaceful international relations system.[37] For the Innovators, traditional diplomacy must be considered in relation to 'the transformed environment of actors, issues and modes of communication within which diplomats function; and yet, demonstrate the continuing centrality of conventional diplomats to most of what happens in contemporary diplomacy.'[38] This approach to writing on diplomacy stresses the on-going importance of the role of traditional diplomacy but in *relation* to emerging forms of non-traditional diplomacy.

4. The Modern Diplomatic Environment

Accepting these three different Schools of thought suggests a far more complex modern diplomatic system than that of states simply dealing with other states. A snapshot of the modern diplomatic environment alludes to this globalised complexity. This snapshot also confirms the limitations inherent to the centuries-old Traditional School as well as the merits of the Nascent and Innovative Schools.

In the modern diplomatic environment the state endures as the dominant political actor while its diplomatic institution (centrally orchestrated by the Ministry of Foreign Affairs) remains the most visible diplomatic actor.[39] Currently, there are 192 states operating in the modern diplomatic environment, compared to 47 in 1950 and 26 in 1926.[40] All of these states interact diplomatically, all need to represent themselves and all need continuously to negotiate advantageous foreign policy ends in a competitive and occasionally hostile environment. With over three-hundred years of related experience, the foreign embassy endures as the leading diplomatic actor, where 'the conduct of relations on a state-to-state basis, via formally accredited resident missions forms the bulk of international exchange.'[41]

But all is not well in the traditional backyard. In the face of ever-tightening budgets and shrinking numbers of overseas missions and staff, the reinvention of the traditional diplomatic institution is apparent.[42] Now these institutions have to justify their relevance to a more informed domestic audience. This justification includes two way communication processes with domestic and global publics (instead of the traditional one way processes which reinforced the hermetic image of many institutions); the inclusion of many non-state 'experts' into the sacred halls of diplomacy; and the shift in recruitment policies to build staffs truly reflective of the domestic societal strata they purport to represent, to name but a few changes. This snapshot suggests that the traditional 'gatekeepers' are struggling to maintain relevance and are reforming in a bid to hang on to the keys to the gate.

On the other side of the gate, energised, increasingly efficient and numerous non-state actors are gathering. There is a correlation between their rising numbers and growing influence. For example, the number of NGOs has risen from 997 in 1954 to 20,928 in 2005-6.[43] The large numbers of MNCs is also noteworthy. At the beginning of the 21st century there are more than 53,000 MNCs, which have over 450,000 foreign affiliates.[44] These ever-growing numbers suggest that MNCs exercise significant clout in the modern diplomatic system, with global assets in excess of $13 trillion (USD) and global sales of more than $9.5 trillion (USD).[45] And more than half of the world's top economies are not countries but global MNCs, with waning affiliations to the nation-state.[46] The growth of inter-governmental organisations (IGOs) in the last hundred years is also significant. In 1909, there were 37 IGOs, by 1962 this number had risen to 163 and by 2005-6 the modern diplomatic environment had 1,963 IGOs.[47] To representatives from this growing non-state sector, the gate no longer looks so imposing or absolute; there are many other paths around the side.

In the modern diplomatic environment, these non-state groups have adopted basic diplomatic functions such as negotiation skills, visible representation, effective communication, filtered information and political reporting from overseas and symbolism (the Greenpeace rainbow flag is instantly recognisable, as are the ubiquitous golden arches). Large MNCs, as one example, are learning of the need to 'develop their own task-defined diplomatic structures to serve their particular needs and develop local expertise that national diplomatic services find hard to rival.'[48]

Where before there was one path and one gate there are now many channels, networks and alternate environments through which to engage in diplomacy. Myriad patterns of asymmetric and polylateral diplomacy are appearing, involving not only state representatives but also representatives from NGOs, transnational organisations (the External Delegations of the European Union for example) and even famous though hardly effective movie stars.[49] The Ottawa Process, the Kimberley Diamonds Process, or the Nazi Gold settlement—success stories of 'unconventional' diplomacy—are regularly trumpeted and are used as evidence to suggest the ushering in of an era of 'new' diplomacy. Whether it is multi-lateralism or summitry or two individuals from different countries conversing on a plane, diplomacy is blossoming and clearly no longer axiomatically linked to the state.[50]

5. The Modern Diplomatic System: Geared for Peace and Stability

An enhanced understanding of diplomacy can only lead to the enhanced practice and application of diplomacy as the business of peace. The notion of diplomacy as the avoidance of conflict and the business of peace is inherent to all three of the Schools, which all interpret diplomacy as the best means for modern civilisation to prevent international relations from being governed by force. An encompassing and increasingly popular definition of diplomacy speaks generally to this consensus. Paul Sharp, from the Innovative School, defines modern, plural diplomacy as 'the way in which relations between groups that regard themselves as separate ought to be conducted if the principle of living in groups is to be retained as good and if unnecessary and unwanted conflict is to have a chance of being avoided.'[51] Significantly, Sharp's definition suggests that modern diplomacy is not dominated exclusively by states and their diplomats, but rather by 'groups.'

Recognising the plural, group nature of the modern diplomatic environment is useful: it compels the diplomatic studies field to substantiate different types of alternate diplomacy, which further adds to the body of knowledge on diplomacy. For example, one area of future research could focus on how an NGO, such as The International Committee of the Red Cross (ICRC), conducts diplomacy. Some initial questions might include: the ICRC operates in the international realm, it has a core interest which it promotes through representation and symbolism and it negotiates with other actors to achieve its policy ends, so the ICRC clearly practices diplomacy but which type? How does it differ from traditional diplomacy, if at all? Is the ICRC more effective at using a form of unconventional diplomacy to tackle unconventional problems? By default, accepting plurality—a notion confirmed by the substantiation of the three Schools of thought—pushes the field towards novel, exciting and untapped areas of research. In other words, recognising the three Schools stimulates a heuristic process of enquiry.[52]

One novel observation on modern diplomacy that the three Schools allude to is the emergence of a different diplomatic system from that which has dominated the international relations system since Westphalia. This ever-thickening diplomatic system—where state and non-state actors co-exist—is driven by rules and norms which are largely peaceful and aimed at mitigating the anarchy inherent to the international relations system.

The modern diplomatic system is largely a state-qua-state system but one that is being increasingly supplemented and complemented by non-state actors. All actors in this system share the common goal of maintaining stability and avoiding conflicts which can damage political relationships, hamper trade or regress the emergence of a global civil society, as but a few examples. As more actors are socialised into this largely peaceful diplomatic system, the merits of solving international problems through diplomacy as opposed to coercion and force become self-evident.

Guiding the pattern of state and non-state interaction is a coherent set of diplomatic rules that each diplomatic actor or group adheres to. These rules are designed to mitigate the complexity of the international relations system through commonality. Different diplomatic actors have different languages, cultures, religions, histories and social values, for example and diplomacy is the common currency running throughout the diplomatic system. In order to reduce friction-leading-to-conflict, diplomacy is governed by a formal and informal 'constitutive framework of principles, rules and organised behaviour in state [and non-state] relations.'[53]

Commonality, familiarity (of rules) and standardisation are central to diplomacy. This type of civil diplomacy is something 'groups' have in common that enables them to communicate in a predictable, familiar, rational, logical and organised manner. State and non-state diplomats are thus socialized as members of a transnational and collegial group with a shared corporate culture, professional language, behavioural codes, entry procedures, socialisation patterns, norms and standards.[54]

The modern practitioners of diplomacy are guided by these standardised norms of interaction. These norms are further endorsed through established diplomatic tools and functions aimed at reducing the propensity for conflict and promoting stability and peace.

These diplomatic functions and tools form the foundations of diplomacy and have their origins buried in the 'dawn of history' with each actor and epoch, including the state and Westphalia, refining and adding to the general practice of diplomacy.[55] Such functions are ubiquitous in the modern diplomatic environment. There are five in total, which were first elucidated by Hedley Bull. First, diplomacy facilitates communication between states and other entities in the international relations system. Without communication 'there could be no international society, nor any international system at all.'[56] Second, diplomacy is the negotiation of agreements, without which 'international relations would be possible but would consist only of fleeting, hostile encounters between one political community and another.' Third, diplomacy is the gathering and dissemination of reliable intelligence and information about foreign actors in the international relations system, which implies a 'constant flow of information about events in all countries…a continuous and universal diplomacy.'[57] Fourth, diplomacy is the minimisation of the effects of friction in international relations. Given the 'juxtaposition of different communities, each with its own values, preoccupations, prejudices and sensibilities, friction in international relations is always present.'[58] It is the task of diplomacy to manage, contain and ease the effects of friction. Fifth, diplomacy fulfils the function of symbolising the existence of an international society, where diplomats 'even in the pristine form of messengers, are visible expressions of the existence of rules to which states and other entities in the international system pay allegiance.'[59]

These five core functions of diplomacy form the bedrock of the activity and are aimed at introducing order, peace and stability to an international relations system beleaguered by anarchy. Given the chance, these functions can lead to a more peaceful, stable and transparent international relations system. Most successfully resolved international problems have been exposed to all of these functions at some stage, whereas acrimonious relationships demonstrate few of the functions described by Bull. With these functions woven throughout the modern diplomatic system, the peaceful nature of the system is further institutionalised.

In the modern sense, diplomacy remains the 'expression of a shared logic of appropriateness' informing the actions of and identities of diplomatic actors while distributing and dispersing shared values, norms and structures to all diplomatic actors.[60] The notions of a shared logic, framework and structure endure at the heart of the modern diplomatic system. As more actors—state and non-state—become part of this system, they adopt the endemic norms, standards and behaviours, which in turn promote dialogue, communication and negotiation as opposed to isolation, miscommunication and non-compliance. All actors then become common stakeholders in the stability of this system, thus further promoting peaceful and diplomatic solutions over potential conflict. Those actors and groups that choose to shun the system or to forge their own diplomatic path to their own policy ends can thus be described as anti-diplomatic and condemned for their willingness to choose conflict and war over peace and stability.

6. Conclusion

This chapter has demonstrated that the modern diplomatic system is more complex than the traditional state-qua-state image suggests. The modern diplomatic system is characterised by polylateral, peaceful and pluralistic networks of diverse diplomatic actors who—for the most part—have an interest in a stable, functioning international relations system. In this modern system, state and non-state actors appear more willing to engage in diplomacy than all out hostility. To further promote this preferable global condition it is important to enhance the understanding of diplomacy, both inside and outside of the diplomacy studies field.

The chapter has shown that just as the practice of diplomacy is increasing in complexity and plurality so too is the study of diplomacy. Introducing, constructing and supporting three different diplomatic Schools of thought illustrates that the scope of diplomacy studies is deepening and broadening. This is a positive evolution, as an enhanced understanding of diplomacy and its many applications can only lead to a more peaceful international relations system.

Notes

[1] R. Cohen, 'Putting Diplomatic Studies on the Map,' *Diplomatic Studies Program Newsletter*, Leicester University, 1998, p. 1.
[2] A.M. Codevilla, 'Tools of Statecraft: Diplomacy and War,' *Foreign Policy Research Institute Footnotes,* January 2008, Viewed on 16 September 2007, <http://www.fpri.org/footnotes/1301.200801.statecraftdiplomacywar.html>.
[3] Cohen, op. cit., p.1.
[4] For a thorough account of the evolution of early diplomacy, see H. Nicolson, *The Evolution of the Diplomatic Method*, Cassell Publishers, London, 1957; R. Cohen, 'Reflections on the New Global Diplomacy: Statecraft 2500 BC to 2000 AD,' *Innovation in Diplomatic Practice*, J. Melisen (ed), Palgrave, London, 1999, pp. 1-18.
[5] P. Sharp, 'For Diplomacy: Representation and the Study of International Relations,' *International Studies Review*, Vol. 1, No.1, 1999, p. 40.
[6] Cohen, op. cit., p. 1.
[7] M. Wight & H. Bull, *Systems of States*, Leicester University Press, Leicester, 1977, p. 113.
[8] To this functional list of diplomacy symbolism and representation might be added.
[9] H. Bull, *The Anarchical Society: A Study of Order in World Politics*, Columbia University Press, New York, 1977, p. 156.
[10] G.R. Berridge, *Diplomatic Theory from Machiavelli to Kissinger*, Palgrave, London, 2001, p. 1.
[11] G.R. Berridge, *Diplomacy: Theory and Practice*, Palgrave, London, 2002, p. 1.
[12] E. Satow, *A Guide to Diplomatic Practice, 4th edition*, Longmans, Green & Co., London, 1957, p. 1.
[13] Ibid.
[14] Ibid.
[15] J. Melissen, *Innovation in Diplomatic Practice*, Maxillon, New York, 1999, p. xvi. The history of the state system suggests that war and diplomacy are not fundamentally opposed; however, this is an area of discussion this chapter will not entertain due to the limit of available space. This chapter argues - simply—that the diplomacy in the 21st century is *largely* the business of peace where warlike behaviour can, at best, be described as anti-diplomatic. There are several scholars that tackle the argument head on. For example, Sofer quotes the younger Moltke after losing the battle of the Marne in 1914 to validate the widely held view that diplomacy, certainly in the past, was employed to induce conflict. Moltke declared that 'the highest art of diplomacy in my opinion does not consist in preserving the peace under all circumstances but in shaping the political situation of the state continually in

such a manner that it is in a position to enter the war under favourable circumstance' See also Fry, Goldstein and Langhorne who note that 'war in the nineteenth century was a normal, legitimate instrument of policy, an extension and logical conclusion of diplomacy' (*Guide to International Relations and Diplomacy*, Continuum, London, 2002, p. ix). Morgenthau also notes the role of diplomacy in the past that 'when war was the normal activity of kings, the task of diplomacy was not to prevent it, but to bring it about at the most propitious moment' (*Politics Among Nations*, Knopf, New York, 1978, p. 592). For a modern discussion on the non-exclusivity of war and diplomacy see S. Joshi, 'Coercive Diplomacy and Weapons of Mass Destruction,' Paper prepared for presentation at the International Studies Association Annual Conference, Montreal, Canada, March 17-20, 2004; A. Stephanson, 'War and Diplomatic History,' *Diplomatic History*, Vol. 25, No. 3, 2001, pp. 393-403; D. Stevenson, 'Militarisation and Diplomacy in Europe before 1914,' *International Security*, Vol. 22, No. 1, 1997, pp. 125-161. And for empirical evidence (mostly from Eastern Europe during the Balkan Crisis) of the inclusivity of force and diplomacy see, C. Bildt, 'Force and Diplomacy,' *Survival*, Vol. 42, No.1, 2000, pp. 141-148. For an opposing argument, that diplomacy and war are mutually exclusive see B. Steiner, 'Preventive Diplomacy in Historical Perspective,' *Diplomacy and Statecraft*, Vol. 9, No. 1, 1998, pp. 1-23.

[16] For the origins and usage of such terms see H. Lacher, 'Putting the State in its Place: the Critique of State-centrism and its Limits,' *Review of International Studies*, Vol. 29, No. 4, 2003, pp. 521-542.

[17] For example, Satow (writing two centuries later) considered the work of De Callieres as a 'mine of political wisdom' (Satow in T.G. Otte, 'A Guide to Diplomacy: The Writings of Sir Ernest Satow,' *Diplomatic Theory from Machiavelli to Kissinger*, G.R. Berridge, M. Keens-Soper & T. Otte (eds), Palgrave, London, 2001, p. 129). Nicolson too expresses admiration for the work of his traditional forefathers, regarding De Callieres' work as 'the best manual on diplomatic method ever written' (*The Evolution of the Diplomatic Method*, Cassell Publishers, London, 1957, p. 62). And the title of a modern, traditional work, Berridge's *Diplomatic Theory from Machiavelli to Kissinger* (2001), indicates an admiration for the earlier scholars writing on diplomacy.

[18] D. Lee & D. Hudson, 'The Old and New Significance of Political Economy in Diplomacy,' *Review of International Studies*, Vol. 30, 2004, p. 345.

[19] A. Watson, *Diplomacy: The Dialogue between States*, Eyre Methuen, London, 1982, p. 12.

[20] F. De Callieres, *The Art of Diplomacy*, M.A. Keens-Soper (ed), Leicester University Press, Leicester, 1983; E. Satow, *A Guide to Diplomatic Practice*, Longman, London, 1957; A. De Wicquefort, *The Ambassador and His Functions*, Leicester University Press, Leicester, 1997.
[21] C.M. Constantinou, 'Late Modern Diplomacies,' *Millennium*, Vol. 22, No. 1, 1993, p. 89.
[22] Watson, op. cit., p. 12. The opinion from Watson and Lee relates to the inadequacy of prescriptive works on diplomacy. Similar rhetoric, relating to such paucity of diplomatic literature, has consensus in the diplomatic studies field. For example, Martin Wight lamented that there is no 'succession of first-rate books about diplomacy' and that few scholars have 'made it their business to study the diplomatic community itself' (M. Wight & H. Butterfield, *Diplomatic Investigation*, Harvard University Press, Cambridge, 1966, pp. 18-22). Similarly, Sofer notes that the 'wealth of literature on diplomacy is quite limited' (Lee & Hudson, op. cit., p. 196). Der Derian argues that, subsequently, the discussion of diplomacy has become 'somewhat marginal to international relations' (*On Diplomacy: A Genealogy of Western Estrangement*, Oxford University Press, Oxford, 1987, p. 3). Neumann concurs when he adds 'overall, the scholarly attention paid to diplomacy is a disappointment' ('The English School on Diplomacy,' Discussion Papers in Diplomacy, Netherlands Institute on International Relations, *Clingendael*, Vol. 79, 2002, p. 8). Before answering their own question, Lee and Hudson ask 'why has diplomacy attracted so little attention?' (op. cit., p. 350). The simple answer is that a statist lens is no longer the only available interpretation of diplomacy. A Traditionalist, partial focus on diplomacy serves only 'to impose a contrived understanding of diplomacy' by 'binding diplomatic identity to a narrow political schema' (Ibid., p. 345-349).
[23] Lee & Hudson, op . cit., p. 354
[24] Two oft cited works challenging this assumption are J. Der Derian, op. cit., and C.M. Constantinou, *On the Way to Diplomacy*, University of Minnesota Press, Minneapolis, 1996.
[25] L. Diamond & J.W. MacDonald, *Multi-Track Diplomacy: A Systems Approach to Peace*, Kumarian Press Books for a World That Works, Virginia, 1996, p. 2.
[26] B. Hocking, *Foreign Ministries: Change and Adaptation*, Macmillan, London, 1999, p. 24.
[27] Der Derian, for example, writes of the 'crisis in which diplomacy finds itself' (op. cit., p. 1) and Riordan of the continuing 'fragmentation of traditional diplomacy' where 'no country, however powerful, will be immune' (op. cit., p. 10). Langhorne adds that 'there has been a decline, since

the First World War, in the role played in international politics by professional diplomacy' (op. cit., p. 2).

[28] This definition was developed, in part, from B. Hocking's 2004 article 'Privatizing Diplomacy,' *International Studies Perspectives*, Vol. 5, pp. 147-152.

[29] This citation was adapted from an article by R. Sil, who was writing on the differences between general theorists in the IR discipline. See R. Sil, 'The Foundations of Eclecticism: The Epistemological Status of Agency, Culture and Structure in Social Theory,' *Journal of Theoretical Politics,* Vol. 12, No. 3, 2000, p. 355.

[30] A body of work reflecting the Innovative School could include: J. Der Derian, *On Diplomacy*, Oxford University Press, Oxford, 1987; B. Hocking (ed), *Foreign Ministries: Change and Adaptation*, Palgrave, London, 1999; J. Melissen (ed), *Innovation in Diplomatic Practice*, Macmillan, New York, 1999; B. Hocking, *Localizing Foreign Policy: Non-Central Governments and Multilayered Diplomacy*, Palgrave, London, 1993; B. Hocking & D. Spence, *Foreign Ministries in the European Union: Integrating Diplomats*, Palgrave, London, 2002; A.F. Cooper, 'Beyond Representation,' *International Journal*, Vol. 53, No. 1, 1997, p. 173; A.F. Cooper & B. Hocking, 'Governments, Non-Governmental Organisations and the Re-calibration of Diplomacy,' *Global Society*, Vol. 14, No. 3, 2000, pp. 361–376; B. Hocking, 'Privatizing Diplomacy?' *International Studies Perspectives*, Vol. 5, 2004, pp. 147-152; P. Sharp, 'For Diplomacy: Representation and the Study of International Relations,' *International Studies Review*, Vol. 1, No. 1, 1999, pp. 33-57; P. Sharp, 'Who Needs Diplomats? The Problem of Diplomatic Representation,' *International Journal*, Vol. 52, No. 4, 1997, pp. 609-634.

[31] A.F. Cooper & B. Hocking, 'Governments, Non-governmental Organisations and the Re-calibration of Diplomacy,' *Global Society*, Vol. 14, No. 3, 2000, p. 361.

[32] Hocking, op. cit., p. 22.

[33] Ibid., p. 23.

[34] One example of this ongoing reappraisal is the paper 'What Challenges Does the Regional Context Present for Diplomacy?,' presented by R. Thakur at the 2006 Diplomatic Update Conference hosted by the Asia-Pacific College of Diplomacy at the Australian National University. In this paper, Thakur innovatively argued that a threefold change in the world of diplomacy and diplomats can be evidenced: first, 'in the levels of diplomacy, from the local through the domestic-national to the bilateral, regional and global; second, in the domain and scope of the subject matter or content of diplomacy; and third, in the rapidly expanding numbers and types of actors'

said to be engaged in diplomacy. This paper is available at: <http://apcd.anu.edu.au/events/2006/diplomaticupdate.htm>.
[35] Melissen, op. cit., p. xx.
[36] M. Kahler, 'Inventing International Relations: International Relations Theory after 1945,' *New Thinking in International Relations Theory*, M. Doyle & G.. Ikenberry (eds), Westview Press, Boulder, 1997, p. 43.
[37] See, for example, P. Sharp's definition of diplomacy, which was introduced earlier in this article under the heading of 'Why Study Diplomacy?'
[38] Sharp, op. cit., p. 47.
[39] For similar usage of the term institution and the reasons behind the employment of this term in place of the Ministry of Foreign Affairs, see C. Jonsson and M. Hall, *Essence of Diplomacy*, Palgrave Macmillan, London, 2005, pp. 39-42.
[40] These figures were obtained from the United Nations, Retrieved 4 February 2006, <http://www.un.org/Overview/unmember.html>.
[41] G.R. Berridge, *Diplomacy: Theory and Practice*, p. 105.
[42] For an account of the recent reform efforts see, B. Hocking, *Foreign Ministries: Change and Adaptation*, Macmillan, London, 1999; K.S. Rana, 'Foreign Ministries: Change and Reform,' Paper presented at the International Studies Association, Montreal, March 18th, 2004.
[43] From Part A: Statistical Data Graphics, Figure 1.1.1. (a): Overview of the number of international organisations by type [types A-G,] of the Yearbook of International Organisations, Edition 42, 2005/2006, edited by the Union of International Associations, Lausanne, Switzerland.
[44] C.W. Kegley Jr. & E.R. Wittkopf, *World Politics: Trend and Transformation, 9th ed.*, Thomson Wadsworth, California, 2003, p. 173.
[45] Ibid.
[46] For example, Exxon-Mobil has a bigger annual economy and revenue than Turkey, Austria, Saudi Arabia and Denmark. For a list of the top 100 economies in the world (countries and companies) see Kegley & Wittkopf, 2003, op. cit., p. 175.
[47] From Part A: Statistical Data Graphics, Figure 1.1.1. (a): Overview of number of international organisations by type [types A-G,] of the Yearbook of International Organisations, Edition 42, 2005/2006, edited by the Union of International Associations, Lausanne, Switzerland.
[48] B. Hocking. 'Privatizing Diplomacy?' op. cit., p. 149.
[49] For an explanation of the formation, purpose and efficacy of these Delegations, see, M. Bruter, op. cit., pp. 183-205.

[50] For a discussion of the phenomenon of people-to-people diplomacy, see P. Sharp, 'Making Sense of Citizen Diplomats,' *Diplomatic Studies Program*, Leicester University, 2001, pp. 1-22.
[51] P. Sharp, 'Herbert Butterfield, the English School and the Civilizing Virtues of Diplomacy,' *International Affairs*, Vol. 79, No. 4, 2003, p. 858.
[52] As Popper and others have pointed out, 'the more we learn about the world and the deeper our learning, the more conscious, specific and articulate will be our knowledge of what we do not know, our knowledge of our ignorance.' Popper offers a useful analogy in relation to the heuristic process inherent to such epistemology: 'It is as though science were working in a great forest of ignorance, making an ever larger circular clearing within which, not to insist on the pun, things become clear...But, as that circle becomes larger and larger, the circumference of contact with ignorance also gets longer and longer. Science learns more and more...We keep, in science, getting a more and more sophisticated view of our ignorance.' K.R. Popper, 'On the Sources of Knowledge and Ignorance,' *Proceedings of the British Academy*, 46, 1960, p. 69.
[53] For example, the formal framework governing relations between states is the Vienna Convention on Diplomatic Relations (VCDR), which came into existence in 1961. The VCDR has several basic functions developed over time: to represent and protect national interests; negotiation; reporting and gathering of information; and the promotion of friendly relations, including economic, cultural, scientific and commercial interests. For discussion on the pre-history and significance of the VCDR, see Richard Langhorne, 'The Regulation of Diplomatic Practice: the Beginnings to the Vienna Convention on Diplomatic Relations, 1961,' *Review of International Studies*, 18, 1992, pp. 3-17. The VCDR is available from the United Nations, retrieved, 4 February 2006, <http://www.un.org/law/ilc/texts/diplomat.htm>. Also see J. Batora, 'Does the European Union Transform the Institution of Diplomacy?' Discussion Papers in Diplomacy, Netherlands Institute of International Relations, *Clingendael*, No. 87, 2003, p. 1.
[54] These ideas were initially suggested by Jozef Batora but refined and developed in this paper.
[55] H. Nicolson, *The Evolution of the Diplomatic Method*, Cassell Publishers, London, 1957, p. 2.
[56] Bull, op. cit., p. 164.
[57] Ibid., p. 165.
[58] Ibid.
[59] Ibid., p. 166.
[60] Ibid., p. 3.

Bibliography

Batora, J., 'Does the European Union Transform the Institution of Diplomacy?' Discussion Papers in Diplomacy, Netherlands Institute of International Relations, *Clingendael*, No. 87, 2003, pp. 1-25.

Berridge, G.R., Keens-Soper, M. & Otte, T. (eds), *Diplomatic Theory from Machiavelli to Kissinger*. Palgrave, London, 2001.

—, *Diplomacy: Theory and Practice*. Palgrave, London, 2002.

Bruter, M., 'Diplomacy without a State: The External Delegations of the European Commission.' *Journal of European Public Policy*. Vol. 6, No. 2, 1999, pp. 183-205.

Bull, H., *The Anarchical Society: A Study of Order in World Politics*. Columbia University Press, New York, 1977.

Cohen, R., 'Putting Diplomatic Studies on the Map.' *Diplomatic Studies Program Newsletter*. Leicester University, 1998.

—, 'Reflections on the New Global Diplomacy: Statecraft 2500 BC to 2000 AD,' in *Innovation in Diplomatic Practice*. Melissen, J. (ed), Macmillan, New York, 1999, pp. 1-18.

Constantinou, C.M., 'Late Modern Diplomacies.' *Millennium*. Vol. 22, No. 1, 1983, pp. 89-96.

—, *On the Way to Diplomacy*. University of Minnesota Press, Minneapolis, 1996.

Cooper, A.F. & Hocking, B., 'Governments, Non-Governmental Organisations and the Re-Calibration of Diplomacy.' *Global Society*. Vol. 14, No. 3, 2000, pp. 361-376.

De Callieres, F., *The Art of Diplomacy*. Leicester University Press, Leicester, 1983.

Der Derian, J., *On Diplomacy: A Genealogy of Western Estrangement*. Oxford University Press, Oxford, 1987.

Diamond, L. & MacDonald, J.W., *Multi-Track Diplomacy: A Systems Approach to Peace*. Kumarian Press Books for a World that Works, Virginia, 1996.

Fry, M.G., Goldstein, E. & Langhorne, R., *Guide to International Relations and Diplomacy*. Continuum, London, 2002.

Hall, M. & Jonsson, C., *Essence of Diplomacy*. Palgrave Macmillan, London, 2005.

Hocking, B., *Foreign Ministries: Change and Adaptation*. Macmillan, London, 1999.

—, 'Privatizing Diplomacy?' *International Studies Perspectives*. Vol. 5, 2004, pp. 147-152.

Kahler, M., 'Inventing International Relations: International Relations Theory after 1945.' *New Thinking in International Relations Theory*. Doyle, M. & Ikenberry, G.J. (eds), Westview Press, Boulder, 1997.

Kegley Jr., C.W. & Wittkopf, E.R., *World Politics: Trend and Transformation*, *9th ed*. Thomson Wadsworth, California, 2003.

Lacher, H., 'Putting the State in Its Place: The Critique of State-Centrism and Its Limits.' *Review of International Studies*. Vol. 29, No. 4, 2003, pp. 521-542.

Langhorne, R., 'The Regulation of Diplomatic Practice: The Beginnings to the Vienna Convention on Diplomatic Relations.' *Review of International Studies*. Vol. 18, 1992, pp. 3-17.

—, 'Diplomacy Beyond the Primacy of the State.' *Diplomatic Studies Program*. No. 43, Leicester University, 1998, pp. 1-11.

Lee, D. & Hudson, D., 'The Old and New Significance of Political Economy in Diplomacy.' *Review of International Studies*, Vol. 30, 2004, pp. 343-360.

Mellisen, J. (ed), *Innovation in Diplomatic Practice*. Macmillan, New York, 1999.

Morgenthau, H.J., *Politics among Nations: The Struggle for Power and Peace, 5th ed.* Knopf, New York, 1978.

Neumann, I.B., 'The English School on Diplomacy.' *Clingendael.* Vol. 79, 2002, pp. 1-28.

Nicolson, H., *The Evolution of the Diplomatic Method.* Cassell Publishers, London, 1957.

Popper, K.R., 'On the Sources of Knowledge and Ignorance.' *Proceedings of the British Academy.* Vol. 46, 1960.

Riordan, S., *The New Diplomacy.* Polity Press, Cambridge, 2003.

Satow, Sir E., *A Guide to Diplomatic Practice, 4th ed.* Longmans, Green & Co, London, 1957.

Sharp, P., 'For Diplomacy: Representation and the Study of International Relations.' *International Studies Review.* Vol. 1, No. 1, 1999, pp. 33-57.

—, 'Making Sense of Citizen Diplomats.' *Diplomatic Studies Program.* Leicester University, 2001.

—, 'Herbert Butterfield, The English School and The Civilizing Virtues of Diplomacy.' *International Affairs.* Vol. 79, No. 4, 2003, pp. 855-878.

Sil, R., 'The Foundations of Eclecticism: The Epistemological Status of Agency, Culture and Structure in Social Theory.' *Journal of Theoretical Politics.* Vol. 12, No. 3, 2000, pp. 353-387.

Sofer, S., 'Guardians of the Practitioners' Virtue: Diplomats at the Warriors Den.' *Diplomacy and Statecraft.* Vol. 16, 2005, pp. 1-12.

The Yearbook of International Organisations, Edition 42. Union of International Associations (ed), Lausanne, Switzerland, 2005/2006.

Watson, A., *Diplomacy: The Dialogue between States.* Eyre Methuen, London, 1982.

Wight, M. & Butterfield, H., *Diplomatic Investigations.* Harvard University Press, Cambridge, 1966.

Wight, M. & Bull, H., *Systems of States*. Leicester University Press, Leicester, 1977.

Stuart Murray is Assistant Professor of International Relations at Bond University, Queensland, Australia.

Part III:

New Wars, History and Cultural Change

The Changing Ottoman Perception of War: From the Foundation of the Empire to Its Disintegration

Mustafa Serdar Palabıyık

Abstract
This chapter provides an account of the evolution of the idea of 'war' in the minds of Ottoman administrators and intellectuals with the aim of revealing how their perceptions of war were transformed in line with domestic and external developments that shaped the empire in their own times. The main argument of the chapter is that there was no uniform Ottoman perception of war; instead, there were particular dominant perceptions for particular periods. While in the initial years of the Empire an understanding of war based on the mobilization of nomadic communities against the Christian 'other,' namely *ghaza*, prevailed, this rather religious perception was gradually replaced with a more mundane conception of war in the mid-16th century onwards. As a result of some significant defeats of Ottoman armies both in the West and in the East starting from the end of the 17th century, it came to be believed that the survival of the Empire could be ensured through diplomacy, not through war. The prevalence of diplomacy over war survived almost until the late 19th century. However, particularly after the end of the Hamidian era (1876-1909) in the first two decades of the 20th century, an idea of resorting to war to save the state developed, which would ultimately result in the Ottoman participation in World War I. In sum, this chapter examines the different representations of war that emerged in different periods of Ottoman history, such as war as a tool of motivation, war as a means of imperial expansion, war as the source of all evils and war as an opportunity to save the state.

Key Words: War, Ottoman Empire, *ghaza*, Katip Çelebi, İbrahim Müteferrika, Abdülhamid II, Enver Paşa, Cemal Paşa.

*****

1. Introduction

From the beginning of the Empire, the Ottomans were not only familiar with the concept of 'war' but even naturalized it as a way of life. In their poems, songs and anthems, they sometimes expressed how they admired the peculiar vehemence and grandeur of war; sometimes how they disliked the destructiveness of its longevity. In other words, war always had a place in Ottoman daily life. Ottomans were either soldiers participating in wars, peasants levied to sustain large armies or commanders and sultans directing the war. Therefore, in order to understand Ottoman society, an examination of the Ottoman perception of war is essential.

This chapter aims to analyse how the concept of war was perceived in the Ottoman Empire from the time of its establishment to its disintegration. Unfortunately, there are limits to the level of detail that can be presented here. It is not possible to examine in detail all conceptions of war advanced by Ottoman statesmen, intellectuals and in popular culture in each period, since the Empire spanned more than six centuries. Therefore, this chapter only emphasizes the most prominent understanding of war in each specific period. It draws attention to how these prominent understandings vary in accordance with the turning points in Ottoman history and correspondingly how the idea of war has evolved in relation to changing internal and external conditions, such as economic problems, social disturbances and/or changing foreign relations and international environments.

The chapter is composed of three sections, which are chronologically sequenced. The first section deals with the rise and fall of the concept of *ghaza* as an ideology of war from the establishment of the Ottoman Empire until early 17th century. Within this framework, exemplified in the chronicles of prominent historians of the period such as Kemalpaşazade, Tursun Bey and Neşri, the differences between a religious understanding of war and its mundane application are illustrated. The second section analyses how Ottoman intellectuals expressed their distress over the longevity of Ottoman wars from the early 17th to the late 18th century and how they embraced the newly emerging 'scientific' understanding of war which argued that wars should be conducted in accordance with 'scientific' principles. In this part, the writings of Katip Çelebi and İbrahim Müteferrika are analysed and two *sulhiyye* (a kind of poem in classical Ottoman poetry on the love of peace) are examined to illustrate the depth of distress concerning war that emerges in Ottoman literature in this period. Finally, the third section deals with the perception of war in the Ottoman Empire during its long 19th century, when the dual processes of modernization and disintegration were co-experienced. In this part, it is shown that the perception of war was transformed from a costly enterprise to be avoided to an opportunity to preserve the Ottoman Empire—as illustrated in the work of 19th century Ottoman statesmen, such as Abdülhamid II, Enver Paşa and Cemal Paşa, as well as Ottoman intellectuals such as Ahmet Cevdet Paşa, Cenap Şehabettin and Namık Kemal. This transformation contributed to the intellectual basis of the decision for Ottoman participation in World War I and thus to the disintegration of the Empire.

2. The Perception of War from the Establishment of the Ottoman Empire to the Treaty of Zsitvatorok (1299-1606)

In the last quarter of the thirteenth century, when the Ottoman *beylik*[1] emerged as a significant power in north-western Anatolia, the Islamic world had recently experienced the dual threat of the Mongols in the east and Crusaders in the west. These antagonistic ventures had consolidated a rather religious perception of war termed *ghaza*, drawing upon the holy elements of Islam. From the largest empires to the smallest political entities, almost all Islamic states of the time embraced this form of warfare as a legitimate use of force outside their sovereign realms.[2]

Although the concept of *ghaza* had a strong religious background, it was widely used for the legitimization of mundane wars fought for the establishment and consolidation of the Ottoman *beylik*, as well as for strengthening its regional superiority. What is more, it was also used to deflect the reaction of surrounding and more powerful rivals in Anatolia and the Mongol-originated Ilkhanid Empire, with whom the Ottomans were still allied at that time. The concept of *ghaza* was also utilized as a tool of motivation among the Turkish tribes in the region.[3] In other words, for directing the energy of settled people along the border with the Byzantine Empire and in particular encouraging nomadic Turcoman tribes to move against the Christian 'other,' *ghaza* served as a powerful ideology.

However, even as early as the very establishment of the Empire, it was not possible to explain every war of the Ottoman *beylik* through the notion of *ghaza*. In particular, the Ottoman struggle with other Muslim political entities of Anatolia accelerated as a result of regional rivalry where non-religious motives persisted. For the Ottomans, even the word *ghaza* was not a simple product of a completely religious/otherworldly perception. For example, the chronicle of 19th century Ottoman historian Mustafa Nuri Paşa (1824-1890),[4] entitled *Netayic-ül Vukuat* (The Consequences of Events), a narrative dialogue between Sultan Murad I (r. 1359-1389) and his grand-vizier Çandarlı Kara Halil Hayreddin Paşa (?-c.1386), noted these non-religious motives embedded in the notion of *ghaza*. Accordingly, Halil Paşa asked the young Sultan under what conditions war became a necessity. The Sultan replied that war became a necessity when there was a just cause and when the expectation of victory was high. Then Halil Paşa said 'that's true, but there is one more condition of war; that is, the interest of the state should be clear in waging war.'[5] In sum, this dialogue demonstrates a very secular understanding of war and shows that the preservation of state interests was a significant factor affecting the perception of war in the earliest stages of the Ottoman Empire. What is more, those who participated in wars also sought material gain, namely slaves or spoils.[6] Even in an anonymous religious text book (*ilmihal*) written in this period, it was stipulated that *ghaza* was a way of acquiring proper livelihood.[7] Thus, *ghaza* was war not only fought for religion but also fought for the profane world.

However, *ghaza* continued to be perceived as a tool of legitimization of the Ottoman wars against the Balkan states and Byzantine Empire until the mid-15th century. As a matter of fact, the battles of Kosovo I (1389), Nicopolis (1396), Varna (1444) and Kosovo II (1448) brought together armies of many European states coordinated by the Papacy; hence Ottomans labelled these battles as 'crusades' and the struggle against the opponents as *ghaza*. The title of one of the anonymous chronicles on the Battle of Varna and some preceding clashes, *Gazavat-ı Sultan Murad bin Sultan Mehemmed Han* (Holy Wars of Sultan Murad, son of Sultan Mehmed Khan), reflected this framing of Ottoman wars as a quest against the 'infidel.'

The notion of *ghaza* continued to shape the perception of war until the conquest of Constantinople; after the conquest, and the greater centralization of the state, campaigns in the western borders of the Empire were brought under more strict state control. Within this framework, the respect for former traditions and the significance attached to *ghaza* decreased to a large extent. For example, 16th century historian Neşrî (?-c.1520) wrote that Sultan Mehmed II the Conqueror (r. 1451-1481) abandoned the tradition of the Sultan's standing in front of his commanders when traditional drums of the army (*nevbet*) were to be beaten.[8] This meant that the Sultan would transform former military traditions peculiar to a *beylik* to ones more suitable for an Empire and an emperor. In this period, *ghaza* was perceived as the ideology of raiders (*akinci*) and dervishes and other anti-centralist elements of the Empire and it gradually lost its importance although it continued to be utilized as a source of legitimization of wars. Until the end of the 16th century, *ghaza*, as an ideology, was left to the raiders in Central Europe and the perception of war in the central imperial army became more political and mundane.

Another significant change in the old war traditions of the Empire after the conquest of Constantinople was the gradual transformation of the soldiers and statesmen of the Empire from the 'friends' or 'comrades' of the Sultan to his 'subjects.' In the establishment of the Ottoman Empire, the Turcoman chieftains such as Samsa Çavuş, Konur Alp and Turgut Alp, recognized the sovereignty of Osman Bey (r. 1299(?)-1326); however, they perceived him as *primus inter pares*, rather than as a master. Raider commanders, who had been pursuing their irregular warfare along the western borders of the Empire, preserved their privileged status during the reigns of Murad I and Bayezid I (r. 1389-1402); however, they began to lose most of their powers with the gradual centralization of the Empire. The relations between raider commanders in the Balkans perpetuating the *ghaza* tradition and the central state deteriorated sharply during the reign of Murad II (r. 1421-1481). For example, raider commanders were held responsible for the defeat of Ottoman army by Serbian-Hungarian troops in the Battle of Zlatica in 1443. In *Gazavat-ı Sultan Murad bin Sultan Mehemmed Han*, it was written that the Sultan had remembered each and every former troublemaking activity of these commanders after this defeat. He then returned to the then capital of the Empire, Edirne (Adrianopolis), in great anger and did not even look at the faces of the raider commanders.[9] This narrative provides an indication of the Sultan's distrust of the raiding mechanism and helps to explain its gradual replacement with the central army.

The third significant matter regarding the transformation of the perception of war during the reign of Mehmed II is the emergence of the idea of 'world dominance,' particularly after the conquest of Constantinople and some other significant conquests of this period, including the territories of the Peloponnesian Greek Kingdom, the Greek-Pontus Kingdom of Trabzon (Trebizond) and many parts of Central and Eastern Anatolia as well as the Crimea. The author of *Tarih-i Ebu'l Feth* (The History of the Father of Conquest) and a contemporary of Mehmed II, Tursun Bey (c. 1426-c. 1490) labelled the Sultan as the 'conqueror of the world' (*sahib-kıran*). He wrote: 'Besides Istanbul, he added twenty countries to the territories inherited from his father and he showed the signs that he was one of the conquerors of the world.'[10] This understanding would be more visible particularly in the reign of Suleiman I the Magnificent (r. 1520-1566).

From the mid-15th to late 16th centuries, it can be said that the motives behind the Ottoman wars can be categorized under three headings.[11] The first set of reasons includes religious motivations. In particular, the battles fought and raids organized in Central Europe were legitimized through religious explanations. A 16th century Ottoman chronicler Kemalpaşazade Ahmed Şemseddin (1468-1533) wrote that by the waging of wars, the Ottoman Sultans served the glory of Islam and acted as the commanders of Muslims.[12] These duties were not only implemented against 'infidels' but also against heterodox sects of Islam, which had been decreed as 'heretical,' and against those states that officially adopted such heterodox sects. For example, when Kemalpaşazade wrote on the Safavids, he said Shah Ismail spread 'the defected creed and evil spirited Shia sect' and thus legitimized the campaign of Selim I (r. 1502-1520) against a Muslim state.[13] Similarly, one of the most significant *şeyhülislam* (a high-ranking religious authority serving as the Grand *Mufti*) of Suleiman's reign, Ebussuud Efendi, was asked whether the massacre of those who belonged to the Shia sect (*kızılbaş taifesi*) could be legitimized on religious grounds, whether those who massacred them could be labelled as *ghazi* and whether those killed by them could be labelled as martyrs (*şehid*). Ebussuud Efendi replied 'in response to that question, yes they were; [killing them] is the highest form of *ghaza* (*ghaza-i ekber*) and [dying at the hands of them] is the highest form of martyrdom.'[14]

A second motive for war involves political considerations. In particular, starting from the third decade of the 16th century, the participation of the Ottoman Empire in the newly emerging European state system and in the alliance structures established among major European states became one of the most significant reasons for the 16th century Ottoman wars. Indeed, as early as the mid-15th century onwards, Italian city states sought Ottoman support in the wars fought among themselves; Ottoman Sultans were aware of the strategic significance of even verbal support. What is more, the Ottoman Empire supported France against the Habsburg Empire, which had attempted to exert control over the entire European continent—the Habsburg Emperor had a similar claim of 'world dominance' as did the Ottoman Sultan. As an expression of support, for example, the Ottoman navy was sent to Toulon in 1543. Meanwhile, through raids organized in Central Europe, the Ottoman Empire contributed to the consolidation of Protestantism in the region against the Catholic Habsburg Empire.[15] In sum, in this period, the Ottoman Empire was an inevitable partner of the European alliance systems because without the Ottoman support, relatively weaker states like England, France and the Netherlands could not effectively cope with the might of the Habsburg Empire. Indeed, the Ottoman Empire struggled with the Habsburgs not only through active military support to its adversaries via alliances, but also through capitulations, which supported the economic development of anti-Habsburg states.[16]

The third set of motives of the Ottoman wars in the 15th and 16th centuries consisted of humanitarian considerations. According to Kemalpaşazade, saving oppressed people from the rule of oppressors was a just cause for war and the Ottoman Sultans assigned themselves the duty of relieving such destitute people from cruel rulers. Kemalpaşazade stressed that one of the reasons for the Rhodes campaign of Suleiman I in 1522 was to save the innocent people of Rhodes from their atrocious rulers.[17]

In sum, in the periods of the establishment and expansion of the Ottoman Empire, the perception of war was transformed. Although the tradition of *ghaza* which had dominated the period of establishment had not totally been abandoned, it was gradually replaced by a more mundane understanding of war. Of course, this does not necessarily mean that religious motives did not influence the perception of war; however, it can be said that while the link between the perception of war and the perception of religion was not totally broken, it was loosened.

3. The Perception of War from the Treaty of Zsitvatorok to the Reign of Selim III (1606-1789)

Towards the end of the 16th century, there emerged indications that the Ottoman Empire's supremacy over the West was beginning to diminish. In the East, the Ottomans engaged in a long battle with the Safavids between 1578 and 1590; whereas in the West they fought with Habsburgs between 1593 and 1606. This meant that the Ottoman Empire was at continuous war with its enemies for almost three decades. Besides the longevity of these wars and the attendant exhaustion of economic resources, the Ottoman Empire began to face technologically better equipped armies, particularly in the West.

The Treaty of Zsitvatorok signed with the Habsburg Empire in 1606 had a symbolic significance. Before this treaty, the Habsburg Emperor was referred to by the Ottomans as the '*Bey* of Vienna,' denoting an inferior status vis-à-vis the Ottoman Sultan. However, in this treaty, the Habsburg Emperor was designated as *Kayzer* (Caesar) meaning, at least legally, that he was perceived as equal to the Ottoman Sultan. In other words, the Ottoman Empire, whose diplomacy was based on the principle of non-recognition of any European state as an equal of the Ottoman Empire and the pursuit of continuous warfare except for temporary truces, began to abandon this strict policy.[18] Still, the Ottoman Sultans perceived themselves as superior to European monarchs, at least at the discursive level, until the late 18th century when permanent embassies were established in several European capitals during the reign of Selim III (r. 1789-1808).

It is evident that Ottoman society and its administrators were weary of these long wars, which neither brought catastrophic defeats, nor significant achievements. Sultan Ahmed I (r. 1603-1617) wrote, for example, in one of his poems how he was distressed over the Ottoman-Habsburg wars of 1593-1606:

> I wanted peace and order; I wanted my soldiers to be in order because this is the world that cannot be inherited by anyone. I wanted peace and tranquillity in the borders; I wanted people to live in peace and prosperity, to organize festivals. I thought about how long this quarrel and clashes would endure.[19]

Besides the psychological pressure stemming from the longevity of the wars, the relative failures of the Ottoman armies in these wars forced 17th century intellectuals to ponder upon the reasons for these setbacks. One of the most prominent intellectuals of this period, Katip Çelebi (1609-1657), cited the deterioration of the order of the Janissary corps, the incapacities of some statesmen and economic difficulties, among others, as the basic reasons for

Ottoman incapacity to achieve decisive victories. However, he also added the negligence of the 'science of geography' as a significant reason for failure. This stipulation was one of the earliest serious warnings about considering scientific methods of war. In his *Tuhfetü'l Kibâr fî Esfari'l Bihâr* (A Gift for Gentlemen on the Sea Campaigns), he emphasized the significance of geography for decisive victories as such:

> It might be difficult to know all aspects of world terrain; however, at least one should know the nature of Ottoman territories as well as adjacent countries, so that if there emerges the necessity to make campaigns and to send troops over there, one could take necessary measures. Taking necessary measures for entering enemy's territory and protecting borderlines becomes much easier with that [the science of geography].[20]

In other words, according to Katip Çelebi, war might be a struggle for the sake of religion; however, this should not preclude the fact that it should be waged in line with scientific principles developed for newer weaponry, strategies and tactics.

The long wars of the 17th century were not only perceived as a matter of distress by Anatolian people, who contributed to these wars both in material terms and in terms of manpower, but also by the esteemed elite of the capital, particularly by the poets. It is not therefore surprising that one of the rarest types of Ottoman classical poetry called *sulhiyye* often appeared towards the end of 17th and beginning of the 18th century. These poems were generally written in the aftermath of a war to praise peace and tranquillity. Two such poems are introduced in this chapter to demonstrate the perception of war in Ottoman literary circles in this period.[21] The first *sulhiyye* was written by one of the most prominent poets of the 17th century, Nâbî (1642-1712), to celebrate the signature of the Treaty of Karlowitz (1699). It addressed the Grand-Vizier of the period, Amcazade Hüseyin Paşa (1644-1702). Although this treaty resulted in significant territorial losses for the Ottoman Empire, in the poem Nâbî perceived it as the 'greatest achievement' and praised the preservation of peace. The poem starts with these lines: 'Thank God that the arena of war and quarrel had closed, the world once more experiences order through peace and tranquillity.'[22] Nâbî also expressed the depressed mood of the Ottoman individual who had been worn down by wars. He wrote:

> No one knew when the noise of war and clashes that had clouded from east to west would end… No one thought of healthiness, the nature of days caught sickness… Neither

religion, nor state had ever experienced such disorder; the realm of Islam had never witnessed such desolation.[23]

In sum, the war had been perceived as a process which brought misery, desolation and chaos in the lines of Nâbî.

The second *sulhiyye* was penned by Sâbit, a 17th century poet, to celebrate the signature of the Treaty of Pasarowitz in 1718. It addressed the Grand-Vizier of the period, Nevşehirli Damat İbrahim Paşa (1662-1730). As with the poem by Nâbî, it also welcomes the end of war and emergence of peace by the grace of God: 'Thank God that with the order of the caliph of Islam the drink of war became prohibited and the drink of peace became permitted.'[24]

These poems show that while holy wars fought for religion and state had been depicted as a festival just two centuries earlier, in the beginning of the 18th century war was perceived as a negative concept to be avoided and peace was praised as an ultimate saviour for the society.

The Treaty of Pasarowitz, which had become a subject of poems praising peace, opened the era called the Tulip Period (1718-1730) in Ottoman history. During this period of temporary peace, military defeats began to trigger modernization movements, albeit very limited ones. The tradition of writing brief but effective pamphlets on the backwardness of the Empire and the reasons for failures in the wars, which had been started in the 17th century by intellectuals such as Katip Çelebi and Koçi Bey (?-c. 1650), continued. In particular, the writings of İbrahim Müteferrika (1674-1745) were of great significance.

Ibrahim Müteferrika was born in 1674 in Colojvar, Hungary, as the son of devout Protestant parents. He was educated in theology at a Protestant monastery until he was 18 years old. Then he participated in the series of wars fought between the Habsburgs and Ottomans between 1683 and 1699 and was captured by the Ottomans in 1692. During his monastery education, he learned Latin and French; after his capture, he added the knowledge of Turkish, Arabic and Persian. Thus, he was educated in the Ottoman Empire as a diplomat and entered into the office of *Reis-ül Küttab*[25] in 1715. Besides his official duty as a diplomat, he was also a renowned philologist, geographer and, most importantly, the first Ottoman printer, since he established the first printing house of the Ottoman Empire belonging to the Muslims in 1727.

His masterpiece, *Usul-ül Hikem fi Nizam-ül Ümem* (The Scientific Methods in the Order of Peoples) was not initially a book, but a report submitted to the Grand Vizier. The purpose of this report was to answer a central question asked by Mahmud I (r. 1730-1754), who was enthroned after the bloody rebellion of a former Janissary officer, Patrona Halil: 'How could those Christians, who had been so weak compared to the Muslims, conquer so many lands in these times and even win over the Ottoman armies that had once been continuously victorious?'[26] The answer given to this question was the lack of knowledge about the enemy. Accordingly, Ibrahim Müteferrika argued that the Muslims should have been informed about their adversaries. They had to be visionary and they had to learn as much as possible of European methods of governance and warfare, European institutions, strategies and tactics. In order to do so, he wrote that he had reviewed both Eastern and Western literature about the former empires and emperors.[27]

Müteferrika also dealt with the reasons for war in his pamphlet, in which he focused on understanding human nature. He wrote:

> The illness of greed, which is present in all human beings from their birth, the inclination to attack other's property, to inflict disorder in another's realm... [These characteristics] are also present in Khans, Sultans and Emperors.[28]

He also mentioned that war was not bound by time; it was a primordial concept valid for all times:

> Ever since the emergence of mankind until our time, it is known that there were wars, clashes, struggles among people and nations on earth. Some nations and states and some of the human beings engaged in bad habits like war in order to obtain worldly interests or to protect themselves from any harm that could come from others.[29]

These lines written by Müteferrika were also in-line with the writings of 17th century British philosopher Thomas Hobbes on human nature. As the aforementioned excerpts indicate, the traces of two of the most important premises of realist international relations theory, namely an essentially evil human nature and the permanence of power struggle in interstate relations, can be found in the writings of Müteferrika. His pamphlet is also important because it generally underestimated religious motives of war, while emphasizing political motives.

The pamphlets and reports of the first half of the 18th century were followed by ambassadorial reports (*sefaretname*) in the second half. Although the establishment of permanent embassies would only be realized at the end of the 18th century, particularly from the 1750s onwards, various envoys had already been sent to European capitals to conclude peace treaties after wars. The ambassadorial reports that emerged out of these missions included analyses of how European armies and states gained supremacy vis-à-vis the Ottoman army and Ottoman Empire. One such ambassadorial report was written by Giritli Ahmet Resmî Efendi (1694-1783) who was sent to Prussia during the years 1763 and 1764. In his report, he fiercely criticized those war-mongering elites who wanted continuous wars as in the old glorious days of the Empire. He wrote that such people were not aware of temporary conditions, the science of war and how to fight; they had just 'dreamed to conquer Rome with a small number of troops.'[30] Ahmed Resmî Efendi also emphasized that determining the foreign policy of the Empire solely on the basis of war was nothing but a ruinous enterprise; he also emphasized the significance of resorting to diplomacy. He accused those who argued for waging war for the survival of the state of ignorance and wanted the administrators not to be influenced by them into declaring war against their adversaries before exhausting all diplomatic efforts.

In sum, during the 17th and 18th centuries Ottoman society perceived war as a negative concept and declared its distress and uneasiness with long and exhausting wars. What is more, in this period, the reasons for continuous defeats were analysed and efforts were made to adopt a scientific approach to war.

4. The Perception of War from the Reign of Selim III until the Disintegration of the Ottoman Empire (1789-1922)

The suspicion of the deterioration of Ottoman institutions in the 17th century resulted in an attitude of caution towards war in the first half of the 18th century. However, the weakness of the Ottoman armies and the collapse of Ottoman institutions were quite evident in the late 18th century. This resulted in the necessity of more serious and radical measures to remedy the weaknesses of the Empire.

The two wars fought with Russia between 1768-1774 and 1787-1791 and the subsequent Treaty of Kuchuk Kainardji (1774) and Treaty of Iassi (1792) clearly demonstrated once more the fragility of the Ottoman Empire. What is more, these long and exhausting wars brought the economic resources of the Empire to the brink of depletion. Bearing another war would be extremely costly. Thus, the defeats of the Ottoman army, which had fought with old methods against well-equipped and well-trained European armies, forced Selim III to take concrete steps to reverse the fortunes of the Empire. In the immediate aftermath of his enthronement, Selim III convened most of the esteemed statesmen of the period and demanded that they prepare reports stating the reasons and the remedies for the deterioration of the Ottoman Empire. Among twenty such reports, thirteen were written by bureaucrats and five of them were written by *ulama* (religious scholars). The military reform proposals included in these reports can be categorized under three categories: (1) the reports prepared by conservatives, which demanded returning to the old customs and methods, laws and regulations of the Suleiman I era; (2) the reports prepared by reformers, which demanded the Janissary corps' adoption of new European methods and technology; and (3) the reports prepared by radicals, which argued that the old army formation could not be reformed, so it should be abandoned and a new army should be established.[31] Among this last group, Koca Yusuf Paşa (1730-1800) proposed the establishment of militia troops from Turkish families living in the provinces of the Empire and the provision of a tax exemption for their families. These troops would be conscripted during times of war. This advice would be realized when Napoleon Bonaparte invaded Egypt in 1798 and for the first time in Ottoman history two regiments were established from Turkish troops in Anatolia with this new method.[32]

The usage of diplomacy was also consolidated during the Selim III period and permanent diplomatic representations were established; at the same time, there were steps taken for the establishment of a European-style army. The *Mühendishane-i Berr-i Humayun* (Royal Military Engineering Academy) had already been established in 1773 and two decades later, in 1793, Selim III opened the *Mühendishane-i Bahr-i Humayun* (Royal Naval Engineering Academy). Such military schools would mushroom in subsequent periods. The *Mekteb-i Ulum-i Harbiye* (War Sciences College) and *Mekteb-i Harbiye* (Military Academy) were founded in 1834 and 1846, respectively. What is more, from European countries, particularly from France, soldiers and military engineers were recruited to modernize the Ottoman army. Even the dress of the soldiers of the new troops, which had been established in accordance with the New Order (*Nizam-ı Cedid*) policy of Selim III, was composed of blue jackets and red trousers inspired by the French army.[33]

The speech that Selim III delivered at his coronation was significant for understanding his thoughts on war and diplomacy. He tried to boost the mood of statesmen and the army by using religious ideas in explaining victory and defeat. According to the Koran, he said, the believers were heralded with victory so long as they met three criteria: abandonment of mundane passions, unconditional obedience to their authorities and abandonment of fear of death because of the honour of martyrdom.[34] Once more, it could be perceived that the religious understanding of war and victory had not totally disappeared from Ottoman discourse. Because of his excitement and devotion, Selim III tried to lead his army in the Ottoman-Russian-Austrian War of 1787-1791. This would have been an extraordinary gesture the likes of which had not been attempted in many years; however, his bureaucrats did not welcome this move and the Sultan was ultimately convinced not to lead the campaign but to assign the command of the Ottoman army to one of his viziers.[35]

While the Ottoman Empire had been recruiting French soldiers and experts to modernize its army and relied on French assistance, it was a shocking experience for Ottoman public opinion as well as for the Sultan when Napoleon Bonaparte, the young and ambitious French general, unexpectedly invaded Egypt in 1798. On the one hand, Selim III wanted to resist this invasion; on the other hand, being aware of Ottoman unpreparedness, he could not declare war on France. He worked to overcome this dilemma. One possibility was establishing alliances with other European states. This would not be a wholly novel departure. Indeed, the history of Ottoman alliance with Christian states could even be traced back to the period of Orhan Bey (r. 1326-1361) when he supported the Byzantine Empire in its struggle against the Serbians in return for the permission to pass through the Dardanelles towards the southern Balkans. In these earlier periods, the Ottoman Empire had always been establishing an alliance with an 'inferior' state for its own interests, which had been declared as a grace to that particular 'inferior' state. However, in the era of Selim III, the alliances made with Russia in 1797 and Britain in 1799 were protective measures taken to ensure maintenance of the Empire, meaning that this time the Ottoman state was the weaker partner and the demanding side.[36] From the invasion of Egypt onwards, the resolution of the problem called the 'Eastern Question'[37] was tied to such alliances.

The first half of the 19th century was also troublesome for the Ottoman Empire. Until 1839 in particular, the Empire was under great pressure. Napoleon's invasion of Egypt was followed by a Serbian uprising in 1804 and a subsequent Ottoman-Russian War between 1806 and 1812. A Greek uprising in 1821 and another Ottoman-Russian War between 1828 and 1829 followed. The uprisings of the Governor of Egypt, Kavalalı Mehmet Ali Paşa and the defeat of the Ottoman army by his troops in Nizip (in south-eastern Anatolia) in 1839 were also quite destructive. Within this period, every war meant a significant loss and a great cost for the Ottoman Empire. By consequence, a strong tendency emerged to pursue diplomacy as a means to ensure the survival of the state. The Ottoman Empire allied with Britain against Russia in 1809, with Russia against Mehmet Ali Paşa in 1833, with France and Britain against Russia in 1854. These alliances contributed to the survival of the state until the 1877-1878 Ottoman-Russian War, after which Britain abandoned its traditional policy of protecting Ottoman territorial integrity. In this century, the periods between 1840 and 1853, 1856 and 1877 and 1878 and 1897 can be described as 'breathing' years for a weak and exhausted empire because there were no large-scale wars.

As far as the perception of war in the 19th century goes, it can be said that the Ottoman intellectuals, particularly the intellectuals of the *Tanzimat* period[38] and the Young Ottoman movement, preferred to stop further deterioration of Ottoman institutions through focusing on the internal problems of the Empire rather than engaging in a belligerent foreign policy to reclaim the lost territories. In other words, they preferred peace rather than war and they tried to realize the political, economic and social reforms necessary to ameliorate the deficiencies of the Empire.

One of the most significant wars of this century was, of course, the Crimean War (1853-1856) because it was the first and only war that the Ottoman Empire fought on the side of France and Britain against Russia. Another significance of this war was that as a result of the emergence of Ottoman newspapers in the mid-19th century, public opinion became more aware of the on-going war, its causes and course. Before the spread of newspapers, people only became aware of wars during the conscription process, the passage of the army from the lands in which they were living or in case of a possibility of an attack on their lands. From the mid-19th century onwards, however, they began to hear quickly about the wars which had been fought in the remotest parts of the Empire. Such awareness also increased the people's moral support of the Ottoman army. For example, Sir Adolphus Slade (1804-1887), who had been serving in the Ottoman navy and who was referred to by the Ottomans as *Müşavir Paşa* (The Advisor Paşa), wrote in his memoirs that the consciousness of the Ottoman people was at the highest pitch during the Crimean War. He admired the courage, motivation and self-sacrifice of ordinary people in Istanbul. What is more, he criticized some of the bureaucrats who had underestimated the power of public opinion during the war.[39]

Ottoman intellectuals began to discuss the concept of 'civilization' in a more systematic way in the second half of the 19th century and this resulted in the evaluation of the concept of war within the framework of these discussions of European civilization. In other words, as in many other fields, so in the field of war, many Ottoman intellectuals argued for the adoption of the European mode of scientific and technological evolution and the perception of war through European lenses. One of the most important intellectuals of this period, Namık Kemal (1840-1888), wrote in the first issue of his newspaper, *İbret*, that he was confident in the future of the Ottoman Empire and among the reasons for his optimism he cited the nature of new Ottoman troops. For him, the courage of the Ottoman troops was still a matter of reality and this courage was one of the main tenets of the survival of the state.[40] What is more, he criticized pessimist views that the Ottoman Empire had nothing with which to thwart its adversaries; rather, he referred to new weapons and naval and military achievements among the instruments for the endurance of the state.[41] However, he still was aware that these achievements were not enough and other precautions were necessary. He argued that the future of the state could be guaranteed not through wishing deteriorating conditions on the adversaries of the Ottoman Empire, but through accumulating more power.[42]

Another prominent Ottoman intellectual emphasizing the detrimental effects of ignoring Western military technology and sciences of war was one of the most esteemed Ottoman bureaucrats and historians of the 19th century Empire, namely Ahmet Cevdet Paşa (1822-1895). He was the official chronicler in the Hamidian era and his chronicle, *Tarih-i Cevdet* (History by Cevdet), provides the reader significant clues to the understanding of war in the second half of the 19th century. One chapter of this chronicle, 'On the Science of War,' summarized the evolution of the science of war (*ilm-i harb*) from the first establishment of regular armies in the period of Philip, the father of Alexander the Great, until the 19th century.[43] Ahmet Cevdet Paşa also explained the regression of the Ottoman Empire through Ottoman failure to adopt Western military technology and so contributed to a long tradition of critique extending from the 18th century onwards. Echoing this tradition, he determined that the victories of Austria, Prussia and France against the Ottoman Empire were a direct result of their advanced military technology.

Ahmet Cevdet Paşa explained the science of war by referring to its two different aspects. Accordingly, the science of war was composed of the sciences of 'strategy' and 'tactics.' Strategy was defined as the determination of suitable places for defence and attack and the placement of the troops in accordance with these specific locations.[44] Tactics, on the other hand, was defined as the organization of the troops and the provision of other measures which would bring victory.[45] According to Cevdet Paşa, as a result of negligence of these two sciences in the Ottoman Empire, from the late 18th century onwards, war was abandoned as a method of inter-state relations and it was replaced by diplomacy. However, this had a negative consequence. Resorting to diplomacy instead of war resulted in a growing disregard of military affairs and the military capabilities of the Ottoman army further deteriorated, which exacerbated Ottoman defeats in the wars.[46] In sum, Ahmet Cevdet Paşa emphasized a vicious circle here. Long wars made diplomacy an essential element of Ottoman foreign policy; this resulted in neglecting military development and when wars did break out after relative periods of peace, Ottoman armies were defeated more severely.

Ahmet Cevdet Paşa perceived war as a conflict of interests and within this framework he developed a compelling realist analysis. His writings exhibited significant differences from the writings of his predecessors. For example, a prominent Ottoman chronicler, Ahmed Vasıf Efendi, (?-1806) wrote in his chronicle, entitled *Mehasinü'l Âsâr ve Hakayikü'l Ahbâr* (The Beauty of Achievements and the Reality of Happenings) and published in 1804, that the strength of European powers stemmed from their religious unity and their perception of the Ottoman Empire as an enemy due to religious considerations. Ahmet Cevdet Paşa, on the other hand, argued that there was no such religious unity in Europe; rather, European peoples adopted different sects of Christianity. Therefore, he stipulated that the policies of European states were determined in accordance with their interests, rather than their religious prejudices.[47]

In two of his writings to Sultan Abdülhamid II (r. 1876-1909), *Tezakir* (Memorandums) and *Maruzat* (Information), Ahmet Cevdet Paşa not only discussed historical events but also their historical ramifications. Abdülhamid II ruled the Empire during the last quarter of the 19th and the first decade of the 20th century and his controversial personality led him to be labelled either positively as the 'Glorious Khan' or negatively as the 'Red Sultan.'[48] Some historians argued that his reign was the last period when a sultan himself fully controlled the state structure; for others, it was simply a period of suppression.

The perception of war during the Hamidian era was not much different from the previous period and the idea of avoiding war and resorting to diplomacy through establishing and switching alliances prevailed. For the Sultan himself, even if the wars ended with victory, their cost was unaffordable to the Ottoman Empire.[49] The Ottoman-Greek War of 1897 was a relevant example of such wars. At the end of the war, although the Ottoman Empire had an evident victory, the intervention of the Great Powers compelled it to cede all the territories it had occupied and to accept a very small war indemnity from Greece. The Greeks, on the other hand, obtained significant concessions despite their defeat, particularly regarding the island of Crete, which would later ease Greek annexation. Experiencing this humiliating victory, Abdülhamid II understood that in order to ensure the survival of the Empire, he had to follow a careful foreign policy. In this way, giving concessions was inevitable and Abdülhamid's motto can be summarized as follows: 'to sacrifice the piece for the sake of the totality.'[50] Ottoman acquiescence to British control of Cyprus in 1878, the French invasion of Tunisia in 1881 and the British invasion of Egypt in 1882 could only be explained by this policy of maintaining Ottoman sovereignty in Anatolia and the Balkans.

The policy of balancing the Great Powers, which was a peculiar characteristic of the Hamidian era, was abandoned with the dethronement of the Sultan in 1909. Between this dethronement and the full usurpation of power by the Committee of Union and Progress (CUP) in 1913, the Ottoman Empire faced two great wars, the Tripoli War of 1911 and the Balkan Wars of 1912-1913. These four years were characterised internally by the continuous power rivalry between the pro-British governments of Kamil Paşa (1833-1913) and pro-German Mahmut Şevket Paşa (1856-1913), who had a great role in ending the Hamidian era and who was supported by the CUP.

The War of Tripoli and the Balkan Wars transformed the perception of war both of the governing elite and in public opinion. This transformation was quite evident in the letters of Enver Paşa (1881-1922), a CUP leader, who was declared as the 'hero of freedom' for his role in the dethronement of Abdülhamid II. Enver Paşa addressed these letters to an unidentified woman, whom he probably met when he served in Germany. The letters sent to this mysterious lady on the eve of Tripoli and Balkan Wars showed a young and ambitious soldier's desire for war.

In this period, young soldiers such as Enver Paşa thought that war was an opportunity to ensure the survival of the state. For example, on August 16, 1912, from his headquarters near Derne, Libya, Enver Paşa wrote that he felt sorrow for not initiating a full-scale war with the Italians. He added that he feared a peace, which would end all his efforts to save the Empire.[51] As understood from these expressions, it was peace that was perceived as a negative concept by the Ottoman soldiers, since every peace brought the Ottoman Empire to the brink of disintegration. This resulted in associating the survival of the state with war. Another letter from Enver Paşa dated June 7, 1913, consisted of only a single sentence: 'Balkans are filled with clouds which are gradually threatening peace; that's so good!'[52] One month after writing these lines, Enver Paşa wrote to the German lady:

> If war would erupt..., this would bring a great advantage for my poor country and for the God's sake we would win...If the disagreement in the Balkans continues, then the war will erupt. This means life for Turkey.[53]

Unfortunately Enver Paşa could not foresee that the Balkan Wars would bring more misery than life to Turkey. Due to lack of communication and coordination among Ottoman troops because of political debates among the commanders of the Ottoman army, the first Balkan War fought between the Balkan states and the Ottoman Empire was lost quickly and all European territories of the Empire right up to the outskirts of the capital were occupied by Balkan armies.[54] However, this time, the Balkan states engaged in another war among themselves to share the spoils of war, allowing the Ottomans to retake some of the lost territories including the former imperial capital, Edirne.

In this period, some of the Ottoman intellectuals began to argue for taking Germany as a model for their own state because of its rapid industrialization and successful modernization. Indeed, from the Hamidian era onwards, German soldiers began to be recruited in the Ottoman army. In particular, Germany's victory over France in 1871 resulted in the replacement of admiration for France with admiration for Germany. One of the most important representatives of the pro-German movement was Cenap Şehabettin. In his travelogue entitled *Avrupa Mektupları* (Letters from Europe), one can easily see his admiration for Germany; this piece also illustrated his perception of war. While he wrote on the military structure of Germany, he mentioned that interstate relations were based on power and emphasized the 'inevitable' supremacy of the strong over the weak.[55]

The perception of war as the only means for survival continued until the eve of World War I and became one of the most significant ideological factors in the Ottoman decision to enter the war. The reflections of this perception can be found in the memoirs, speeches and writings of major decision-makers of that period. For example, at a dinner gathering of the members of the Triumvirate and other prominent CUP members,[56] Cemal Paşa (1872-1922), the commander of the Ottoman Fourth Army based in Syria, explained the reason why the Ottoman Empire had entered the war as follows: 'Gentlemen, if the Ottoman Empire had not participated in war, the independence of the country would have been in a grave danger.'[57] At the same dinner, Halil Bey (1874-1948), the speaker of the Ottoman Parliament, emphasized that the Ottoman Empire could only maintain its territorial integrity through participating in the war. What is more, Ottoman victory, for Halil Bey, would not only save the Ottoman Empire but also all Muslims and Turks living outside the borders of the Empire.[58] Enver Paşa made a similar comment in saying that they would not only succeed in the protection of the Ottoman throne but also ensure the survival of the whole Islamic community.[59] As these expressions demonstrate, World War I was the epitome of the Ottoman perception of war as an opportunity to save the Empire.

Another significant issue regarding the Ottoman participation in World War I was the declaration of jihad by the announcement of the text entitled 'The Holy Jihad Declaration' (*cihad-ı mukaddes ilanı*), which was signed by *şeyhülislam* Ürgüplü Hayri Efendi (1867-1921) and approved by Sultan Mehmed V (r. 1909-1918) on November 14, 1914. This text was translated into many languages spoken in Muslim countries and sent to almost all the Asian and African states where Muslims had been living. What is more, in order to attract the support of the Shia community, the Shia religious leaders of the holy cities of Najaf, Karbala and Kazemiye were commanded to inform the people of the legitimacy of participation in World War I on the side of the Ottoman Empire.[60] Thus, the Ottoman Empire hopelessly tried to attract all Muslim material and human resources into World War I; however, it failed since most of the Muslim communities were under control of colonial powers and their sons fought in colonial armies against the Ottoman Empire. This also showed the limits of caliphate in the 20th century.

5. Conclusion

This chapter aimed to analyse the perception of 'war' in Turkey from the establishment of the Ottoman Empire to its disintegration. In doing so, this study has tried to reveal how the perception of war was transformed in line with the domestic and external developments experienced by the Empire. In the final analysis, it can be said that there was no uniform perception of war; rather, there were some particular dominant perceptions for particular periods. In the initial years of the Empire, in order to direct the energy of nomadic tribes towards the Christian 'other,' the perception of *ghaza* prevailed, despite the fact that even in these years this 'other' sometimes appeared as the ally of the Ottomans. However, with the centralization of the Empire, particularly from the mid-15th century onwards, religious dimensions of the understanding of war were replaced by a more secular understanding, perceiving war as a mundane issue. Although, in this period, Ottoman Sultans perceived themselves as superior to their European or Asian counterparts, as a result of long and tiresome wars of the 17th century and subsequent Ottoman defeats, war had been defined as a process between equal rulers.

During the 18th century, attempts were made to remedy Ottoman defeats through development of a more scientific understanding of war. The need to do so is particularly evident in the writings of intellectuals such as Katip Çelebi and İbrahim Müteferrika. The 19th century, on the other hand, was a period in which war had been discredited and diplomacy was perceived to be the most viable tool for the survival of the state. Despite these ideas favouring avoidance of war, towards the end of this century and particularly in the first decade of the 20th century, the Young Turks, who had dethroned Abdülhamid II and re-established a parliamentary regime, adopted the idea of 'war for the survival of state.' This would result in the Ottoman participation in World War I and the subsequent disintegration of the Empire.

All in all, war was always a part of Ottoman society along with statecraft despite the fact that there was no uniform perception of it. Rather, depending on domestic and external circumstances, the perception of war varied. War was perceived as a tool of motivation, as a means of expansion, as a devilish game to be avoided and as an opportunity for the survival of state.

Notes

[1] A *beylik* was an administrative unit which denoted the political entities formed in Anatolia after the disintegration of the Seljukid Empire in the late thirteenth century. Among many *Beylik*, the Ottoman Empire was initally one of the smallest ones based in northwestern Anatolia, in the territories adjacent to the Byzantine Empire. The administrator of a *Beylik* was called the '*Bey.*'

[2] The word *ghaza* stems from the word *ghazva* connoting the wars in which the Prophet Mohammed himself participated. In its narrowest sense, it meant 'fighting with the enemy.' In the late 13th and early 14th centuries this notion meant the spread of Islam and enlargement of the territories controlled by Muslims. After 1260, Çobanogulları *Beylik* was assigned to fight with the

Byzantines and among the labels used for Çobanogulları was *nusret-ul guzzat* (meaning the victor of *ghaza*). The label *ghazi* (meaning the one fighting for God) was also used for denoting the *Bey* of Sinop and Aydın, who were contemporaries of Osman Bey, the founder of the Ottoman Empire. C. Kafadar, 'Gaza,' *Islam Ansiklopedisi*, Vol. 13, 1996, p. 427.

[3] Ibid., p. 427.

[4] The dates between parentheses indicate the dates of birth and death; while the dates with a mark of 'r.' indicate the periods of reign for the Sultans.

[5] M.N. Paşa, *Netayic-ül Vukuat*, Neşet Çağatay (trans & ed), 4 Volumes, Türk Tarih Kurumu Yayınları, Ankara, 1995, Vol. 1, p. 278.

[6] Kafadar, op. cit., p. 428.

[7] Ibid.

[8] C. Kafadar, 'İki Cihan Aresinde,' *Doğu-Batı, Special Issue on Ottomans,* no. 19, 1999, pp. 41-61.

[9] M. Cezar (ed), *Mufassal Osmanlı Tarihi*, 6 Volumes, İskit Kitabevi, İstanbul, 1958, Vol. 1, p. 311.

[10] R. Murphey, 'Fatih Sultan Mehmet Döneminde Osmanlı İç ve Dış Siyaseti,' *Osmanlı*, G. Eren (ed), 12 Volumes, Yeni Türkiye Yayınları, Ankara, 1999, Vol. 1, pp. 239-246.

[11] Kemalpaşazade, *Tevarih-i Al-i Osman*, Ş. Severcan (ed), Türk Tarih Kurumu Yayınları, Ankara, 1996, p. LVI.

[12] Ibid.

[13] Ibid. p. LIX.

[14] E. Düzdağ, *Şeyhülislam Ebussud Efendi'nin Fetvaları Işığında 16: Asır Türk Hayatı*, Enderun Kitabevi, İstanbul, 1972, p. 109.

[15] For a detailed analysis of Ottoman-European relations in the early modern period, see Mustafa Serdar Palabıyık, *Contributions of the Ottoman Empire to the Construction of Modern Europe*, Unpublished M.Sc. Thesis, Middle East Technical University, Institute of Social Sciences, Ankara, 2005.

[16] The capitulations were granted to France in 1569, to England in 1580 and to the Netherlands in 1612.

[17] Kemalpaşazade, op. cit., p. LXII.

[18] A.İ. Savaş, 'Genel Hatlarıyla Osmanlı Diplomasisi,' *Osmanlı*, G. Eren (ed), Vol. 1, pp. 643-659, esp. p. 644.

[19] Quoted from M.N. Paşa, *Netayic-ül Vukuat*, p. 204.

[20] O.Ş. Gökyay, *Katip Çelebi'den Seçmeler*, Milli Eğitim Bakanlığı Yayınları, Ankara, 1968, p. 114.

[21] For these poems see A.F. Bilkan, 'İki Sulhiyye Işığında Osmanlı Toplumunda Barış Özlemi,' *Türkler*, H.C. Güzel, et al. (eds), 21 Volumes, Yeni Türkiye Yayınları, Ankara, 2002, Vol. 12, pp. 598-605.

[22] Ibid., p. 599.

[23] Ibid., pp. 600-601.
[24] Ibid., p. 603.
[25] *Reis-ül Küttab*, literally the 'chief of clerks,' assumed the duty of administering daily routines of Ottoman foreign relations from the Treaty of Karlowitz onwards.
[26] İ. Müteferrika, *Milletlerin Düzeninde İlmi Usuller*, Ö. Okutan (ed), based on the original text that had been published in 1732 by the printing house of İbrahim Müteferrika, Milli Eğitim Bakanlığı Yayınları, İstanbul, 1999, p. 7.
[27] It is important to note that İbrahim Müteferrika could read and write in six different languages: Hungarian, French, Latin, Turkish, Arabic and Persian. Thus it is not surprising that he could even quote ancient Greek philosophers such as Socrates, Aristotle and Democritos in his book.
[28] Müteferrika, op. cit., p. 23.
[29] Ibid., p. 33
[30] S. Çataltepe, 'III. Selim Devri Askeri Islahatı ve Nizam-ı Cedid Ordusu,' *Osmanlı*, Eren (ed), Vol. 7, pp. 241-249.
[31] N. Berkes, *Türkiye'de Çağdaşlaşma*, Yapı Kredi Yayınları, İstanbul, 2002, p. 98.
[32] Ibid., p. 102.
[33] H. Erdemir, 'Batılılaşma Sürecinde Fransa Etkisi,' Güzel et al. (eds), *Türkler*, Vol. 14, pp. 641-646.
[34] Quoted from E.Z. Karal, *III. Selim'in Hatt-ı Humayunları*, Türk Tarih Kurumu Yayınları, Ankara, 1999, p. 23.
[35] Ibid., p. 24.
[36] Savaş, op. cit., p. 655.
[37] Indeed, the Eastern Question is a very complex phenomenon; however, it can be briefly defined as the international rivalry for domination over the Ottoman territories from the late eighteenth century until the early twentieth century. In other words, the concept of the 'Eastern Question' does not refer to a particular problem, rather a variety of issues emerged out of the Ottoman decline.
[38] This period labelled as the Reordering Period started with the imperial edict declared in 1839, in which several reforms regarding the political, economic and social structures of the Empire were proclaimed.
[39] B. Özcan, 'Kırım Savaşı (1853-1856),' *Osmanlı*, Eren (ed), Vol. 2, pp. 97-112.
[40] 'İbret,' No. 1, 13 June 1872, *Namık Kemal ve İbret Gazetesi*, M.N. Özön (ed), Yapı Kredi Yayınları, İstanbul, 1997, p. 48.
[41] 'İbret,' No. 3, 17 June 1872, Ibid., p. 56.
[42] 'İbret,' No. 2, 15 June 1872, Ibid., p. 52.

[43] A.C. Paşa, *Tarih-i Cevdet*, 3 Volumes, Üçdal Neşriyat, İstanbul, 1976, Vol. 1, pp. 167-175.
[44] Ibid., p. 172.
[45] Ibid., p. 172.
[46] Ibid., p. 173.
[47] C. Neumann, 'Araç Tarih Amaç Tanzimat: Tarih-i Cevdet', *Siyasi Anlamı*, Tarih Vakfı Yurt Yayınları, İstanbul, 2000, p. 143.
[48] See, for example, A.T. Kutlu, *II. Abdülhamid: Kızıl Sultan Mı, Ulu Hakan Mı?*, Nokta Yayınları, Istanbul, 2007.
[49] E. Akarlı, 'II. Abdülhamid: Hayatı ve İktidarı,' *Osmanlı*, Eren (ed), Vol. 2, pp. 253-274.
[50] B. Kodaman, 'II. Abdülhamid Hakkında Bazı Düşünceler,' *Osmanlı*, Eren (ed), Vol. 2, pp. 275-285.
[51] Ş. Hanioğlu (ed), *Kendi Mektuplarında Enver Paşa*, Der Yayınları, İstanbul, 1989, p. 165.
[52] Ibid., p. 245.
[53] Ibid., p. 245.
[54] A.T. Alkan, 'Meşrutiyetten Cumhuriyete Ordu-Siyaset İlişkileri,' *Türkler*, Güzel, et al. (eds), Vol. 14, pp. 765-781, esp. p. 774.
[55] C. Şehabettin, *Avrupa Mektupları*, T.C. Kültür Bakanlığı Yayınları, Ankara, 1995, p. 39.
[56] After 1913, Enver Paşa, Cemal Paşa and Talat Paşa assumed the control of government; hence, their rule was termed as Triumvirate.
[57] H. Babacan, 'Enver Paşa,' *Türkler*, Güzel et al. (eds), Vol. 13, pp. 263-273, esp. p. 269.
[58] Ibid., p. 270.
[59] Ibid.
[60] S. Sarısaman, 'I. Dünya Savaşı'nda Türk Cephelerinde Psikolojik Harp,' *Türkler*, Güzel, et al. (eds), Vol. 13, pp. 453-468, esp. p. 458.

Bibliography

Ahmet, C.P., *Tarih-i Cevdet*. 3 Volumes, Üçdal Neşriyat, İstanbul, 1976.

Akarlı, E., 'II. Abdülhamid: Hayatı ve İktidarı.' *Osmanlı*. Vol. 2, Eren, G. (ed) Yeni Türkiye Yayınları, Ankara, 1999.

Alkan, A.T., 'Meşrutiyetten Cumhuriyete Ordu-Siyaset İlişkileri.' *Türkler*. Vol. 14, Güzel, H.C. et al. (eds), Yeni Türkiye Yayınları, Ankara, 2002.

Babacan, H., 'Enver Paşa.' *Türkler*. Vol. 13, Güzel, H.C. et. al. (eds), Yeni Türkiye Yayınları, Ankara, 2002.

Berkes, N., *Türkiye'de Çağdaşlaşma*. Yapı Kredi Yayınları, İstanbul, 2002.

Bilkan, A.F., 'İki Sulhiyye Işığında Osmanlı Toplumunda Barış Özlemi.' *Türkler*. Vol. 12, Güzel, H.C. et al. (ed), Yeni Türkiye Yayınları, Ankara, 2002.

Çataltepe, S., 'III. Selim Devri Askeri Islahatı ve Nizam-ı Cedid Ordusu.' *Osmanlı*. Vol. 7, Eren, G. (ed), Yeni Türkiye Yayınları, Ankara, 1999.

Cezar, M. (ed), *Mufassal Osmanlı Tarihi*. 6 Volumes, İskit Kitabevi, İstanbul, 1958.

Düzdağ, E., *Şeyhülislam Ebussud Efendi'nin Fetvaları Işığında 16: Asır Türk Hayatı*. Enderun Kitabevi, İstanbul, 1972.

Erdemir, H., 'Batılılaşma Sürecinde Fransa Etkisi.' *Türkler*. Vol. 14, Güzel, H.C. et al. (eds), Yeni Türkiye Yayınları, Ankara, 2002.

Gökyay, O.Ş., *Katip Çelebi'den Seçmeler*. Milli Eğitim Bakanlığı Yayınları, Ankara, 1968.

Hanioğlu Ş. (ed), *Kendi Mektuplarında Enver Paşa*. Der Yayınları, İstanbul, 1989.

İbrahim, M., *Milletlerin Düzeninde İlmi Usuller*. Milli Eğitim Bakanlığı Yayınları, İstanbul, 1999.

Kafadar, C., 'Gaza.' *Islam Ansiklopedisi*. Türkiye Diyanet Vakfı Yayınları, İstanbul, 1996, Vol. 13, pp. 427-428.

—, 'İki Cihan Aresinde.' *Doğu-Batı, Special Issue on Ottomans*. No. 19, 1999, pp. 41-61.

Karal III, E.Z. *Selim'in Hatt-ı Humayunları*. Türk Tarih Kurumu Yayınları, Ankara, 1999.

Kemalpaşazade, *Tevarih-i Al-i Osman*. Severcan, Ş. (ed), Türk Tarih Kurumu Yayınları, Ankara, 1996.

Kodaman, B., 'II. Abdülhamid Hakkında Bazı Düşünceler.' *Osmanlı*. Vol. 2, Eren, G. (ed), Yeni Türkiye Yayınları, Ankara, 1999.

Kutlu, A.T., *II. Abdülhamid: Kızıl Sultan Mı, Ulu Hakan Mı?* Nokta Yayınları, İstanbul, 2007.

Murphey, R., 'Fatih Sultan Mehmet Döneminde Osmanlı İç ve Dış Siyaseti.' *Osmanlı*. Vol. 1, Eren, G. (ed), Yeni Türkiye Yayınları, Ankara, 1999.

Mustafa, N.P., *Netayic-ül Vukuat*. Türk Tarih Kurumu Yayınları, Ankara, 1995.

Neumann, C., *Araç Tarih Amaç Tanzimat: Tarih-i Cevdet'in Siyasi Anlamı*. Tarih Vakfı Yurt Yayınları, İstanbul, 2000.

Özcan, B., 'Kırım Savaşı (1853-1856).' *Osmanlı*. Vol. 2, Eren, G. (ed), Yeni Türkiye Yayınları, Ankara, 1999.

Özön, M.N. (ed), *Namık Kemal ve İbret Gazetesi*. Yapı Kredi Yayınları, İstanbul, 1997.

Palabıyık, M.S., *Contributions of the Ottoman Empire to the Construction of Modern Europe*. Unpublished Thesis, Middle East Technical University, Institute of Social Sciences, Ankara, 2005.

Sarısaman, S., 'I. Dünya Savaşı'nda Türk Cephelerinde Psikolojik Harp.' *Türkler*. Vol. 13, Güzel, H.C. et al. (eds), Yeni Türkiye Yayınları, Ankara, 2002.

Savaş, A.İ., 'Genel Hatlarıyla Osmanlı Diplomasisi.' *Osmanlı*. Vol. 1, Eren, G. (ed), Yeni Türkiye Yayınları, Ankara, 1999.

Şehabettin, C., *Avrupa Mektupları*. T. C. Kültür Bakanlığı Yayınları, Ankara, 1995.

Mustafa Serdar Palabıyık received a PhD degree in International Relations from the Middle East Technical University and is currently teaching at the University of Economy and Technology in Ankara. He specializes in Ottoman diplomatic history and Turkish foreign policy, with a particular emphasis on the Turkish perceptions of the Orient/Occident debate in the late nineteenth and early twentieth centuries.

Youth and Peace: Alternative Voices in Lebanon

Pamela Chrabieh Badine

Abstract
In the last few decades, Lebanon has witnessed several armed conflicts, piling new war memories onto old, reviving old wounds and shattering the dreams of reconstruction and stability. In this context of multilayered social and political crises and conflicted identities, the role of Lebanese civil society, Diaspora and transnational civil society in spreading courage and hope despite constraining conditions, and in working towards breaking the cycle of War by building a culture of Peace is of the utmost importance. This chapter presents a summary of recent research conducted on five organizations (including youth groups and a collection of forty Lebanese individuals between 25 and 40 years old). In their work and output, these groups and individuals illustrate the necessity of testifying to personal war experiences, learning the lessons of history, sharing the mourning, breaking the cycle of hatred and revenge, connecting divisions and embracing a common humanity. In doing so, they run against the grain of the political class and large parts of Lebanese society, forming a cross-sectarian 'counter-culture.'

Key Words: Lebanon, memory of war, peace, interreligious dialogue, civil society, youth.

*****

In the last four decades, Lebanon has witnessed several conflicts divided by brief periods of tense peace which have led to more than 150,000 casualties, 190,000 injured, 13,000 kidnapped, 18,000 missing, massive destruction, millions of displaced persons, billions in debt, to name only the most evident effects. Various internal and external tensions and conflicts intertwined together led to the breakdown of the Lebanese government and public institutions and the outbreak of war. Nevertheless, the majority of commentaries explain the causes of war as religious or sectarian or as foreign invasions and interferences in local affairs.[1] Even Islamic-Christian Dialogue groups and movements in Lebanon perceive the causes of war to be interreligious.

On these perspectives, the focus is either on issues of violence and particularly the relation between religion and violence or on Islamic-Christian coexistence. The plurality of discourses and spaces of peace that do not emphasize these issues are disregarded or marginalized. This is the case for voices-paths for peace especially in large parts of the youth, women and

Diaspora communities, whose contributions are sharply undervalued. Yet, since the nineties—and even before—the emergence of discourses and practices within the civil society and the Diaspora reveals the existence of new political forces (alternative forces to the traditional ones) and an enriching diversity in the production of knowledge concerning war and peace issues—a diversity that is not confined to religious and political leaders and elites.[2]

Having conducted field research to study these forces and this diversity in the last decade, I discovered that young Lebanese men and women between 25 and 40 years of age constitute the largest and most effective age group engaged at a personal or a collective level in rethinking these issues, after surviving the 1975-1990 period as children and/or teenagers.[3] These individuals coped with their traumas after being deeply troubled for years and constructed personal and collective war memories towards dialogue and conviviality. This chapter introduces the reader to some of their experiences, visions and contributions.

1. Research Methodology

This research shows how young Lebanese understand, imagine and view reconstruction in the Lebanese context as well as by changing socio-political and cultural dynamics. Therefore, interdisciplinary and qualitative methodologies were used. The former, applied along with a conceptual approach at the crossroads of Social Sciences, Science of Religion, Political Science and Cultural Studies, enables the researcher to study the cultural phenomenon of the Lebanese youth by integrating it in the complexity of its social, political and historical contexts. Furthermore, it allows the confrontation of a 'fragmented knowledge' in order to constitute a configuration that complies with the particular cognitive expectations and interrogations of the research.

The first phase of the research took place in Montreal and in Beirut from September 2005 to May 2006. It consisted of content analysis and participation in a number of events and meetings organized by young Lebanese. The method of participant observation was used in a process of dynamic immersion until the field research was finalized in 2007. I focused particularly on five groups and associations: first, *Nahwa al-Muwatiniya* (Towards Citizenship), a non-profit organization founded by young individuals in Lebanon in 2005 looking to build a democratic society. Second, I engaged with the *Art of Living Foundation*, the largest volunteer based Non-Governmental Organization in the world. In Lebanon, amongst its diverse initiatives, Art of Living organizes trauma relief and Peacebuilding projects. Third, I worked with *Tadamon* (Solidarity), a new Montreal-based collective of social justice organizers and media activists, working to build relationships of solidarity with grassroots political movements for social and economic justice between Montreal and Beirut. Fourth, I participated in the activities of *Helem* (Dream), a group previously known as *Club Free*. Helem has been working on LGBT (Lesbian, Gay, Bisexual and Transgender) issues in Lebanon in recent years (as well as in Ottawa, Montreal, Sydney, Paris and San Francisco). Its activities have included lobbying with other human rights organizations for the advancement of human rights and personal freedoms in Lebanon. Fifth and finally, *Le Cercle Interreligieux de Montréal*, a Montreal based group of Islamic-Christian Dialogue, was founded in 2003. This group of 15-20 permanent members, including Lebanese activists organized Peacebuilding projects and conferences about War and Reconciliation in Lebanon in 2004 and 2005. Moreover, it produced and animated several Radio and TV programs (Radio Canada, 'Les guerriers repentis,' Second Regard, 2005, Radio Ville-Marie, Canal Vox, etc.).

Given the lack of academic data concerning the subject of my research, it was also necessary to interview forty Lebanese individuals with diverse confessional or non-confessional identities (25% Christians, 20% Muslims, 15% Atheists, 30% Agnostics, 5% believers with no institutional affiliation, 5% private religious beliefs and practices), political and socio-economical identities, while using a qualitative semi-directive method. Each interview was preceded by the filling out of a questionnaire in order to collect the basic information about the identity of the interviewee. An equal number of men and women were chosen in order to allow the often marginalized discourses and practices of women in Lebanon on all levels (social, political, economical, religious) to be recognized.[4] 50% of the interviewees identified themselves as Lebanese, while 40% identified as Lebanese and Canadian and 10% as Lebanese and 'other' (French, Brazilian, etc.). The research focused on Lebanese living in Lebanon and in Canada—especially in Montreal, where more than 66% of the Lebanese in Canada are settled. Among the 40% of the Lebanese-Canadian interviewees, nearly 30% live in Lebanon and 70% in Montreal while regularly visiting Lebanon and working in local and transnational organizations to help their country of origin.

The majority of the interviewees were involved in a group, association or non-governmental organization (80% local, 75% transnational) concerned with Human Rights (50%), Social Justice-Caritative (30%), Interreligious and Intercultural Dialogue (10%) and Peacebuilding (50%): these included *Amam 05, Arab Resource Collective, Association Lotedhal, Fédération des Femmes du Québec (FFQ), Focolari (Liban), Gais et Lesbiennes Asiatiques de Montréal (GLAM), Gris, Groupe de recherche et d'intervention sociale (Montréal), Haya Bina, Injaz, International Gay and Lesbian Youth Organization, Kafa, Lebanese Student Association (Concordia, Montréal), Lebanus (Canada), Nahnou, Ou3a, Palestiniens et Juifs Unis (PAJU), Penal Reform International, Queer Peace International, Raoul Follereau, Rassemblement Canadien pour le Liban, Réseau Euromed des Droits de l'Homme, Rotary Club (Sahel-Metn, Liban), Sawa Group, Solida (Soutien aux Libanais Détenus Arbitrairement), Union Internationale de la Presse Francophone* and *Youth Association of the Blind.* Beginning in September 2005, the interviewees were asked numerous questions including the following: how do you define the war in Lebanon? How do you remember and memorialize the war (or how do you construct personal war memories)? How do you use your personal war memories to define your identity? Do you find it significant to construct-reconstruct a national war memory as part of a national reconstruction process? How do you link your personal memories to the construction-reconstruction of this national memory? How do you link the construction-reconstruction of a national war memory to other pending issues such as the Religious-Political-Social relations and especially, the issue of Sectarianism? What are the strategies-discourses-actions addressing these issues that you develop and promote on the individual and collective levels in Lebanon and Canada?

2. The War in Lebanon: A Diversity of Definitions

Building a Culture of Peace necessitates a rethinking of the causes and impacts of the Culture of War in Lebanon and especially an analysis of the war representations-visions within the Lebanese civil society, Diaspora and transnational civil society.[5] These representations-visions appear to be diversified, whether in the academic field or the popular and media fields: 'civil war,' 'Islamic-Christian war,' 'war of the others,' 'war for the others,' 'social and economic war,' etc.[6] This diversity is reflected in the data collected within our research: less than 5% of the interviewees identify the war as a 'civil war' or an 'Islamic-Christian war,' confirming the thesis of Lebanese analysts such as Carole Dagher; and 5% of the interviewees identify the war as a 'war of others' or a 'war for others,' echoing the thesis of the prominent journalist Ghassan Tuéni. For these two categories of interviewees, the war started in 1975 and ended in 1990. Nevertheless, a third category of interviewees (90%) identifies the war as 'continuous' and 'plural'—in other words, the war in Lebanon has no precise beginning, no ending and has multiple forms: civil, sectarian, social and economical, a war of clans, families and tribes, a war of generations, a war for the others, a war of others on Lebanese soil, etc. Here are some examples of their testimonies:

> The war in Lebanon does not have a single purpose and therefore has many faces. The regional conflict fueled existing internal differences that derive from the race between the eighteen communities to govern the country.

> I think that the war in Lebanon has several aspects. It began with supporters of the Arabic camp and those of the Western camp. It degenerated into a civil war.
>
> The war has not ended yet. Lebanon has never found sovereignty or independence.
>
> The war has not yet finished (...); only deliberate large-scale anarchic violence has stopped.
>
> I think that the socio-political factors of the war existed since the rise of the first Republic. 1975-1990 was an episode of bloody consequences and the factors still exist.
>
> Lebanon survives on doses of morphine. It is a fertile ground for violence that could explode at any moment
>
> The war is not over, even if 'they' tell us so. It continues even if there is no military fighting and destruction of property. Its causes are many and date back to decades of accumulation of injustice and inequality.
>
> The war is multifaceted: civil, religious, between social classes war, war of others.
>
> The war is multifaceted. But this is not an interfaith war. Not fought, but ideas and affiliations. It is a war of others and a civil war (Lebanese-Palestinian, Lebanese-Syrian, Lebanese-Israeli, 'Inter-Lebanese,' etc.). In my opinion, it is a mixture of mini-wars that followed one another, constantly fed by the warlords.

According to these interviewees, the war in Lebanon is a 'blend' of an 'invisible war' (symbolic, psychic, a war of representations-visions) run by 'passive actors' who fail to understand the nature of their complex memories and identities and its counterpart, the 'visible war' (physical—combats, negotiations, treaties) run by 'active fighters.'[7] Ending the war necessitates breaking the vicious circle of the 'visible-invisible,' and/or working on both the 'political' and the 'human,' the 'material' and what the NGOs identify as the 'non-material dynamics' (traumas, suffering, marginalization, injustice).

3. Constructing the Peace Process

Ending the war compels all Lebanese to undertake Peacekeeping (the enforced prevention of further violence) and Peacemaking (the forging of an actual settlement) processes targeting the 'active fighters,' as well as a Peacebuilding process targeting both 'active fighters' and 'passive actors' with the purpose to build Conviviality between the diverse identities and memories within the Lebanese civil society, Diaspora and transnational civil society. In fact, in the Lebanese context, what I call 'the three Ps' (Peacekeeping, Peacemaking and Peacebuilding) are interrelated processes, requiring both structural and relational transformations—mainly between non-combatants (or passive actors) and with ex-active fighters. All social actors are and have to be included in the Peace process.

According to most of my interviewees, the Peace process in Lebanon requires several strategies—patterns, broad statements, long term plans of action that set a direction—including Conflict Transformation (Organizational Leadership/Mediation-Negotiation), Human Security Development and Violence Prevention, as well as Trauma Healing and Reconciliation, to name only the most evident. In order for the last two strategies to be implemented, here are three major steps to follow:

- A first step is to recognize the plurality of the personal and collective memories of the war.
- A second step is to build a national-transnational memory of the war based on these personal and collective memories.
- A third step is to build bridges of conviviality between the Lebanese, based for the most part on this national-transnational memory of the war.

A. Recognizing the Personal and Collective Memories of the War

Most of the interviewees have started to construct memories of the war based on their personal perceptions and experiences since the mid- or late-nineties, but very few researches have been devoted to the study of these memories. Several investigations of the geo-political, military and economic aspects, as well as studies of traditional collective memories (political parties, confessions, militias) have been conducted but very little has been done to present a map of the discursive terrain of non-state actors or has been devoted to the way Lebanese from all backgrounds perceive the war.

Personal perceptions of the war have been depicted in fiction, memoirs, diaries, music productions, paintings and novels throughout the eighties and nineties—and even after—by Lebanese from all backgrounds and identities, relating to all who are concerned with human survival, depicting the horrors of the war, the agony of individuals and families and their struggles to keep their souls alive.[8] They have also been portrayed in documentaries and movies since the late nineties. Ziad Doueiri's film (*West Beirut,* 1998) cleverly interweaves the facts with the emotions of ordinary people in the Lebanese capital torn apart by war in 1975. Josef Fares presents in his film (*Zozo*, 2005) the story of a boy, Zozo, who is living in Beirut during the war. Fares depicts the horrors of the war and the way it affects civilians. A similar depiction is presented in Khalil Joreige and Joana Hadjithomas's film (*A Perfect Day,* 2005). This film reveals the agony of the families who have to choose between waiting or moving on.

These depictions and many more express a particular type of memory based on contrition and solidarity facing the war, commitments in favour of peace and reconciliation, courage and hope despite constraining conditions.[9] They are part of a popular culture which is solely 'allowed' to open Lebanon's wartime files—especially what is not revealed by the Government, the political parties and the warlords. Even in this cultural arena, frank discussion of the war arouses considerable controversy.

However, the individuals I interviewed seemed to be more concerned with telling the truth than being afraid of its impact and more concerned with their generation's diversities (Christians, Muslims, Atheists, Agnostics, Buddhists, Hindus, Lebanese, Lebanese-Canadian, Humanists, Leftists, Feminists, Gays, Lesbians and even what Patrice Brodeur designates as 'Glocal' identities, referring to newly emerging forms of religiosity or spirituality that transcend the boundaries of religious identities) than the older generations, recognizing the plurality of war representations and experiences.[10]

Most of the individuals and the groups I have contacted for my research and most of the individuals I interviewed are more concerned with rethinking the past in order to build a better future than reducing their present to a quest for survival and thus shrinking the self to its minimal dimensions. In their discourses, 'living' and operating a self-criticism appears to be more important than 'surviving' and being trapped in the 'victim state of mind'—without marginalizing the status of the victims and the responsibilities for the violence that were overlooked by the Amnesty Law passed in 1991. Indeed, this law was passed to put an end to the hostilities and achieve some type of stability. However, it effectively excused most Lebanese citizens and officials from prosecution for war crimes and crimes against humanity committed between 1975 and 1990.

Moreover, these individuals and groups often use informal, alternative or even underground strategies of testifying, rather than traditional demand-making actions:[11] publishing articles and independent media reports on the Internet, launching alternative websites and blogs, producing alternative radio programs, short-movies, documentaries, trauma plays or dramas of survival, organizing home movie screening events as well as artistic experimental exhibitions (e.g., Zico House in Hamra Street in Beirut), round tables and debates using 'new media' resources (video, interactive installation, Internet and virtual reality) and mixing literary and artistic genres. These strategies are promoted mostly by decentralized, interactive and intercultural systems such as the Lebanese Blogosphere. A decentralized system has no clear leader, no hierarchy and no headquarters. It is an open system in which everyone is entitled to make her/his own decisions. In other words, the power lies with each individual and is distributed across geographic spaces.[12]

Since July 2006, I have identified more than eighty blogs for young Lebanese in their twenties and thirties living in Lebanon and abroad, fighting war through the internet. Most of these bloggers illustrate old invasions and massacres through their personal 'souvenirs.' Here are some examples of my interviewees' testimonies:

> I remember that the school administration forced us to follow a leader and we went by bus to participate to a demonstration. We were children!
>
> I used to teach in a school. The same speech we used to hear as children was promoted: 'We Christians, You Muslims, We Muslims, You Christians', 'We followers of a party advocating for the Lebanese nation and You traitors' (…). I thought I was living in 1989-90.
>
> I was three or four years old. My father was kidnapped (…). I remember being on the balcony with cousins and seeing a soldier—a militiaman arresting him. Fortunately, he was back forty-eight hours later, but this incident traumatized me.
>
> I remember the bombing, bunkers, snipers, the quest for survival, death—but also the will to live and cope with violence and also more recently the assistance to the displaced following the war of summer 2006.

Bloggers also comment on recent combats, using anecdotes, photographs, video footage and drawings, while showing that the construction-reconstruction of a national-transnational memory of the war begins with the representation of wartime events lived by 'ordinary people' and the promotion of social conditions that are conducive to creativity and innovation. The rise of the Lebanese Blogosphere during the Summer of 2006 combat not only constitutes a response-resistance to the Israeli Invasion, but also creates alternative localities (*lieux alternatifs*) for a diversity of Voices-Paths to be communicated, expressed, debated and shared, thus redefining the political sphere while accelerating the crossing of all kinds of boundaries (national, religious, political, cultural) and the promotion of a unified, pluralistic and mediated society. These localities reveal that social and political spaces are never structured neutrally like a *tabula rasa*, but instead consist of individual and collective discourses and practices, visions and interpretations of discursive forces. They show that the structures of the Internet and new media offer a range of possibilities for subjective expressions and that in times of limited mobility, individuals excluded from traditional spaces have found a way to exercise their autonomy.

The Lebanese Blogosphere Culture is a part of a Culture of Resistance that is not ephemeral—it has been evolving since the mid-nineties within the Lebanese civil society, Diaspora and transnational civil society. The resistance focuses not only on survival and is not driven by hatred or revenge—even if the testimonies of some individuals I interviewed during the first three weeks of the Israeli invasion in 2006 focused on its horrific impacts and many expressed their anger and fear without restraint. 'Bomb Tel Aviv' was a common despair-inspired reaction amongst some young individuals, whether Muslims, Christians or Atheists.

This alternative to the traditional culture promoted in Lebanon by the political parties and leaders provides a multilevel process for dealing with the regional conflicts and the national-transnational problems while targeting the implementation of a self-regulating society with unlimited access to information, where people actively participate in determining their destiny. On the regional level, this culture aims not only at ending the conflicts, but also at breaking the status quo, by trying to develop alternatives to successive failed peace processes that are not convincing anymore to the Lebanese and Arab populations as a management practice during intervals between conflicts. On the national level, it does not simply aim at implementing a long-term cease-fire. It aspires to break the internal status quo.

B. Building a National-Transnational Memory of the War

The second step calls for the construction-reconstruction of a national memory of war. For some of the interviewees, there is no national memory of the war in Lebanon, but only individual and collective memories. A national memory must then be constructed. For others, a national memory of the war exists, but is the memory of the leaders—the warlords—the militias, the traditional parties, the traditional religious institutions and the mainstream media. In that sense, the Lebanese find themselves with this memory—the memory of power, *la mémoire du pouvoir*—as the only national-transnational memory of the war that is transmitted from generation to generation. This memory needs to be deconstructed and reconstructed on the basis of the diverse individual and collective memories—and especially the memories emerging from the 'intimate spaces of culture' to be found locally and abroad.[13] In April 2008, hundreds of Lebanese marched through Beirut to commemorate the 33rd anniversary of what is usually identified as the outbreak of the war in 1975. One of the activities accompanying the march was a pause of the participants in the Tayyouneh roundabout, where black and white pictures of the Lebanese who went missing in the 1975-1990 period were on display. This is an example of an important statement from non-state actors against war and for personal-collective memories and commemorations to be included in the construction-reconstruction of a national memory.

In other words, there should be an ongoing process of interaction between individual and collective readings of the diverse pasts and presents—a plural and dynamic process to which the constructions-representations-expressions of all actors of the Lebanese civil society, Diaspora and transnational civil society contribute. The main objective would be to go beyond old and recent wounds and concentrate on the most important duties-responsibilities as national, diasporic and transnational citizens: the exploration and understanding of similarities and differences, dialogue, reconciliation and conviviality. These duties involve justice, recognition and healing. It is about helping each other move forward with a better understanding of the deep echoes of the past. It is about providing a strong sense of how we define ourselves in our societies and how we take responsibility for solving our own problems.

C. Building Bridges of Conviviality between all Lebanese Actors

Finally, constructing or reconstructing a national-transnational memory of the war (and therefore, renewing the Lebanese identity) calls for building up a long-term conviviality between the diversity of memories and identities—a conviviality that is not only interreligious or inter-sectarian as the mainstream media tended to portray the Relief Campaigns and social solidarity in July and August 2006, but is also between generations, nationalities, social classes and genders. An inter-human conviviality in all its dimensions can be generated through these individuals and collectivities, at least as a starting point. That is why the idea of a national-transnational dialogue is being promoted, a dialogue that is not reduced to political and religious leaders. This dialogue should integrate private sector leaders, academics, technocrats and young activists, whose expertise would help balance the needs of all Lebanese citizens and allow Lebanon to tackle the significant challenges it faces in areas like sectarianism, environmental degradation, massive youth emigration, debt, unemployment, corruption, feudality, public sector inefficiency, Palestinian refugees camps and shortcomings in urban and rural planning, to name only the most obvious.

The inter-human conviviality not only calls for dialogue but also for a reconciliation process that could be implemented by reinforcing the already existent initiatives undertaken by individuals and collectivities in Lebanon and abroad. It could also be implemented by applying the strategies listed below which were identified by the interviewees: the creation of online forums for dialogue amongst individuals, collectivities and Lebanese state representatives; the creation of spaces of hospitality, understanding and reconciliation within schools, universities, municipalities, youth groups, religious communities, private and public firms, Media, etc.; the creation of a committee to rewrite the Lebanese contemporary history books for schools and universities—a committee which would include experts, elites and other diverse actors of the Lebanese civil society, Diaspora and the transnational civil society—or to revive and reform the Civic Education and History Programs which were elaborated in 1997-2001 by the *Education Center for Research and Development* in Lebanon; the creation of academic departments and research centers on War and Peace; and the construction of commemorative spaces on the basis of partnerships between a diversity of individuals and collectivities (War museums, exhibitions on War and Peace in schools and universities in Lebanon and abroad amongst the Lebanese Diaspora). These spaces should tackle the individual and collective memories of the war and not only those of the leaders and elites. Also, it is important to identify a national day to commemorate the victims of all wars in Lebanon.

Two other strategies were also identified: the establishment of a national-transnational media campaign for Peace and the creation of a Truth and Reconciliation (TRC) Commission. The first should involve both mainstream media and alternative independent media. This campaign should raise awareness about the need for a backwards looking review (war crimes, human rights abuses), while covering stories of the victims more effectively. It should also inform local people of the dangers of wars and their negative impacts by publishing pamphlets and showing slots on television.

As for the creation of a TRC commission, it should be adapted to the Lebanese history and context. It should coordinate initiatives to give assistance to victims of wars and their families. For example, a national network of Trauma Clinics would provide victims of war with a professional counseling service and outreach programs in order to allow them to overcome their ordeals. Regardless of what other forms of compensation could be offered, it is essential that those who suffered themselves—through torture or assault, or who lost family members through assassination or disappearance—are given access to Trauma Clinics and professional counseling services that will allow them deal with the pain of the past. Self-help groups could be created, consisting of either victims or the families of victims that could share common experiences and solutions and establish support structures between victims or the families of victims. This may prove particularly valuable in circumstances where people are feeling alienated and isolated from the broader community. Also, an essential component of building reconciliation and healing the wounds of the past is to offer those who participated in violent acts an opportunity to discuss their motives as well as their fears. It is fundamentally important that this 'violence debriefing' be offered to all perpetrators of war and human rights abuses, particularly to those who have been granted amnesty. Lebanon's never-ending war has yet to undergo such unflinching public scrutiny.

4. Future Prospects

In order for the peace process to be constructed, many obstacles need to be overcome: for example, the weak application of the law; *Sultawiyyah* (Hegemony) and *Ahadiyyah* (Homogeny) which have been the currency of rule; the latent state of violence which prevents any peace process from being undertaken on a large scale; the political and religious divisions which split the forces of the Lebanese population; the absence of cooperation and consensus which prevents the implementation of innovative projects; the profusion of scattered powers and foreign interventions in national affairs; the despair that affects many individuals and collectivities within the Lebanese civil society, Diaspora and the transnational civil society; the marginalization of the study of peace issues and implementation by the state, the traditional political parties, leaders and media; the phenomenon of 'Ivory Towers'—or the lack of long-term partnerships and cooperation—amongst most of the NGOs and the Youth groups-associations (except for a few attempts).

When the collective assets within the Lebanese civil society, Diaspora and transnational civil society are more coordinated and mobilized as pluralistic national-transnational assets, their effectiveness and value as problem-solving tools grows exponentially. Expertise and resources can be specifically shared transnationally in order not to duplicate efforts where these can be scarce. As no single individual can act alone to meet the needs of the entire Lebanese population, no group-association acting alone can address today's national-transnational challenges. Therefore, there is a constant need for multilateralism through effective cooperation and partnerships, beginning with individual initiatives and extending from the most basic unit of socialization to the most expansive relationship between citizens, collectivities and the state. This national-transnational perspective can be useful in developing common standards that can later make joint action more efficient and effective.

With all the surrounding difficulties and obstacles, the culture of peace in Lebanon in the decade ahead resides in the ongoing endeavour of individuals and collectivities to act for marginalized voices, to re-craft the relations among themselves and with others, to work on implementing a long-term 'positive Peace' process and thus not only to try living or surviving in the absence of criminal or political violence—'negative Peace.' Lebanon's Reconstruction will surely benefit from this vision and its application on a large scale, leading to the bankruptcy of the war ideologies and culture and of the sentimental pathos accompanied by an occultation of the problems nurturing the war in Lebanon. Finally, let me end this chapter with a quote of a young Lebanese activist following the July 2006 conflict:

> We share with all youth a desire for peace, democracy and sustainable development (…). Lebanese youth refuse to live in the past. We refuse to survive in fear. We refuse to be robbed of our future. The 'New Middle East' has to be a place where we work together to learn from the lessons of history (…).The 'New Middle East' has to be a place where we respect our borders with one another and break the vicious cycle of hatred and revenge with tolerance and forgiveness. It has to be a place where we embrace our common humanity.[14]

Notes

[1] Refer to: G. Tueni, *Une guerre pour les autres.* Jean-Claude Lattès, Paris, 1985; K. Salibi, *A House of Many Mansions: The History of Lebanon Reconsidered*, University of California Press, Berkeley, 1990; F. Traboulsi, *A History of Modern Lebanon*, Pluto Press, London, 2007.

[2] In my research the concept of civil society includes all citizens who participate individually and collectively in the social and political dynamics of the country (except state employees), whether residing in Lebanon or abroad. As for the concept of Diaspora, its use refers to a diversity of definitions of exile and it moves between the specificity of an historical experience to an existential condition.

[3] This post-doctoral research is entitled 'Voices-Paths of Peace in Lebanon. Contributions of the 25-40 Lebanese Age Group to the National Reconstruction.' The first phase took place in Montreal and in Beirut from September 2005 to May 2006. It was financed by the Canada Research Chair of Islam, Pluralism and Globalization (Université de Montréal). The second phase took place at the same Chair and at the Institute of Islamic-Christian Studies (Université Saint-Joseph de Beyrouth, Lebanon) from June 2006 to May 2008. This phase was financed by the Social Sciences and Humanities Research Council of Canada (Government of Canada). The results of the research were published in book format by *Dar el-Machreq* (Université Saint-Joseph de Beyrouth, Lebanon, 2008).

[4] Refer to: M. Cooke, *War's Other Voices: Women Writers on the Lebanese Civil War*, Cambridge University Press, Cambridge, 1998; N. Nauphal, 'Post-War Lebanon: Women and other War-Affected groups,' *ILO Action Programme on Skills and Entrepreneurship Training for Countries Emerging from Armed Conflict,* 1997, <http://www.ilo.org>.
[5] The concept of transnational civil society refers to the system/web of national NGOs, individuals and transnational groups-associations-organizations that interact at the transnational level with each other, with governments or with intergovernmental and semi-governmental organizations (eg. United Nations, World Bank, etc.).
[6] B. Labaki, K. Abou-Rjeily, *Bilan des guerres du Liban, 1975-1990*, L'Harmattan, Paris, 1993; L.F. Tarazi, *An Occasion for War: Civil Conflict in Lebanon and Damascus*, The Centre for Lebanese Studies & I. B. Tauris Publishers, London, 1994; L. Abul-Husn, *The Lebanese Conflict: Looking Inward*, Lynne Rienner Publishers, Boulder, London, 1998; G. Corm, *Le Proche-Orient éclaté 1956-2003*, Folio Histoire, Paris, 2003, pp. 1005-1014.
[7] Refer to: A. Houballah, *Le virus de la violence*, Albin Michel, Paris, 1996.
[8] Refer to: M. Labaki, *L'enfant du Liban. Les chemins de nulle part*, Fayard, Paris, 1986; E. Accad, *Coquelicot du massacre*, L'Harmattan, Paris, 1988; S. Bechara, *Résistante*, Jean-Claude Lattès, Paris, 2000; J.S. Makdissi, *Beirut Fragments: A War Memory*, Persea Books, New York, 1990.
[9] Refer to: A. Messarra (ed), *Monitoring Civil Peace and Memory in Lebanon (From the Remembrance of War to a Peace Culture)*, Lebanese Foundation for Permanent Civil Peace, Ayia Napa Conference Centre, Cyprus and Konrad Adenauer Foundation, Librairie Orientale, Beirut, 2004.
[10] Refer to: P. Brodeur, 'From Postmodernism to Glocalism: Towards a Theoretical Understanding of Contemporary Arab Muslim Constructions of Religious Others,' *Globalization and the Muslim World*, B. Schaebler & L. Stenberg (eds), Syracuse University Press, New York, 2004, pp. 188-205.
[11] For more details about traditional demand-making actions, refer to: M. Dawahare, *Civil Society and Lebanon: Toward a Hermeneutic Theory of the Public Sphere in Comparative Studies*, Universal Publishers, Macquarie Park NSW, 2000; S. Amin & A. Kenz, *Le monde arabe: Enjeux sociaux, perspectives méditerranéennes*, L'Harmattan, Paris, 2003; S. Khalaf, *Cultural Resistance: Global and Local Encounters in the Middle East*, Saqi Books, London, 2002; K. Karam, *Le mouvement civil au Liban: Revendications, protestations et mobilisations associatives dans l'après-guerre*, Karthala, Paris, 2006.
[12] Refer to: O. Brafman & R.A. Beckstrom, *The Starfish and the Spider: The Unstoppable Power of Leaderless Organizations*, Portfolio, New York, 2006.

[13] Refer to: M. Herzfeld, *Cultural Intimacy: Social Poetics in the Nation-State*, Routledge, New York, 2005, p. 12.
[14] Refer to: R. Merhi, 'Lebanese Youth Call on US to Respect Democracy,' *The Daily Star*, Beirut, September 8, 2006.

Bibliography

Abul-Husn, L., *The Lebanese Conflict: Looking Inward.* Lynne Rienner Publishers, Boulder, London, 1998.

Accad, E., *Coquelicot du massacre.* L'Harmattan, Paris, 1988.

Amin, S. & Kenz, A., *Le monde arabe: Enjeux sociaux, perspectives méditerranéennes.* L'Harmattan, Paris, 2003.

Bechara, S., *Résistante.* Jean-Claude Lattès, Paris, 2000.

Brafman, O. & Beckstrom, R.A., *The Starfish and the Spider: The Unstoppable Power of Leaderless Organizations.* Portfolio, New York, 2006.

Brodeur, P., 'From Postmodernism to Glocalism: Towards a Theoretical Understanding of Contemporary Arab Muslim Constructions of Religious Others'. *Globalization and the Muslim World.* Schaebler, B. & Stenberg, L. (eds), Syracuse University Press, New York, 2004.

Cooke, M., *War's Other Voices: Women Writers on the Lebanese Civil War.* Cambridge University Press, Cambridge, 1998.

Corm, G., *Le Proche-Orient éclaté 1956-2003.* Folio Histoire, Paris, 2003.

Dawahare, M., *Civil Society and Lebanon: Toward a Hermeneutic Theory of the Public Sphere in Comparative Studies.* Universal Publishers, Macquarie Park NSW, 2000.

Fawaz Tarazi, L., *An Occasion for War: Civil Conflict in Lebanon and Damascus.* The Centre for Lebanese Studies & I. B. Tauris Publishers, London, 1994.

Herzfeld, M., *Cultural Intimacy: Social Poetics in the Nation-State.* Routledge, New York, 2005.

Houballah, A., *Le virus de la violence.* Albin Michel, Paris, 1996.

Karam, K., *Le mouvement civil au Liban: Revendications, protestations et mobilisations associatives dans l'après-guerre.* Karthala, Paris, 2006.

Khalaf, S., *Cultural Resistance: Global and Local Encounters in the Middle East.* Saqi Books, London, 2002.

Labaki, B. & Abou-Rjeily, K., *Bilan des guerres du Liban, 1975-1990.* L'Harmattan, Paris, 1993.

Labaki, M., *L'enfant du Liban. Les chemins de nulle part.* Fayard, Paris, 1986.

Makdissi, J.S., *Beirut Fragments: A War Memory.* Persea Books, New York, 1990.

Merhi, R., 'Lebanese Youth Call on US to Respect Democracy'. *The Daily Star.* Beirut, September 8, 2006.

Messarra, A. (ed), *Monitoring Civil Peace and Memory in Lebanon (From the Remembrance of War to a Peace Culture).* Lebanese Foundation for Permanent Civil Peace, Ayia Napa Conference Centre, Cyprus and Konrad Adenauer Foundation, Librairie Orientale, Beirut, 2004.

Nauphal, N., 'Post-War Lebanon: Women and Other War-Affected Groups'. *ILO Action Programme on Skills and Entrepreneurship Training for Countries Emerging from Armed Conflict,* 1997. Accessed September 16, 2008, <http://www.ilo.org>.

Salibi, K., *A House of Many Mansions: The History of Lebanon Reconsidered.* University of California Press, Berkeley, 1990.

Traboulsi, F., *A History of Modern Lebanon.* Pluto Press, London, 2007.

Tueni, G. *Une guerre pour les autres.* Jean-Claude Lattès, Paris, 1985.

Pamela Chrabieh Badine holds a PhD in Sciences of Religions (Université de Montréal, Québec, Canada). She is Associate Research Scholar and the Director of International Relations at the Canada Research Chair in Islam, Pluralism and Globalization (Université de Montréal). She is also a

Researcher and a Teaching Fellow at the Institute of Islamic-Christian Studies (Université Saint-Joseph de Beyrouth, Lebanon). She published numerous scientific articles in Canada and Lebanon on the Interreligious Dialogue, the Politics-Religions-Society relations in the Near Eastern area and on the role of the Lebanese civil society and Diaspora in building Peace. She also published the following books: *Icônes du Liban, au carrefour du dialogue des cultures,* Carte Blanche, Montréal, 2003; *À la rencontre de l'Islam, itinéraire d'une spiritualité composite et engagée*, Médiaspaul, Montréal, 2006; *Voix-es de paix au Liban, Contributions de jeunes de 25-40 à la reconstruction nationale*, Dar al-Machreq, Beyrouth, 2008; *Quelle gestion des diversités au Liban? Du confessionnalisme au pluralisme médiateur*, Editions Universitaires Européennes, Sarrebruck, Allemagne, 2010; and *La gestion de la diversité au Liban: Visions de jeunes du secondaire*, Dar el-Machreq, Beyrouth, collection 'Interaction islamo-chrétienne'- Institut d'études islamo-chrétiennes de l'Université Saint-Joseph de Beyrouth, 2009.

The Correspondent's Experience of War

Tim Markham

Abstract
This chapter takes a political phenomenological approach to the study of war, with the aim of explaining the key determinants of individuals' conscious experience—in this case, war reporters. This perspective avoids both the voluntarist conception of consciousness as 'owned' by social agents and the determinism of rational actor theory, holding instead that consciousness is structured by a variety of factors and in sometimes unpredictable ways. Such determinations can be interpreted as strategic, in the sense that individuals and institutions have an interest in how symbolic forms such as professional values and identities are structured. The chapter draws on a discourse analysis of interviews with war correspondents and others active in the field, including military personnel and PR officers, newspaper editors and a US government spokesperson. It identifies two symbolic economies which govern the symbolic value accorded to individuals and media forms, esotericisation and ambivalence, and suggests that while these mostly go unacknowledged there is some scope for reflexivity amongst war reporters concerning the 'game' in which they are engaged. The rules of interaction proceed according to distinctions made between actors, usually not overtly but through expressions of common sense or personal character. However, these rules are subject to change. The chapter assesses how the physical constraints of pool and embedded reporting and the imposition of foreign scheduling influence a journalist's ability to perform his/her professionalism and autonomy and ends by suggesting four ways in which new communication technologies produce substantially different experiences of inhabiting a conflict environment.

Key Words: War reporting, journalism, political phenomenology, Pierre Bourdieu, field theory, communications technology.

*****

1. Introduction

The aim of this chapter is to assess the professional-political determinants of the war reporter's experience of conflict situations. Traditional studies seeking to explain actors' behaviour in war have, unsurprisingly, concentrated on military and political personnel. These have largely used Clausewitzian or Hobbesian macro-conceptions of political economy,[1] though more recent work has developed a more 'soldier-centred' method,[2] in line with parallel trends in other areas of the social sciences which have sought to locate the conscious reflections of individuals at the heart of sociological research, over a more structurally deterministic framework. In the conference panel entitled *Theoretical Philosophy in the Analysis of War,* Jeff and Bill Harmon presented interesting research from experimental psychology which seeks to avoid both the reductive functionalism of institutional/organisational analysis as well as the individualistic voluntarism of ethnography. The authors examine the determinants of behaviour at what they call the 'genetic' level: in particular, how physiological and endocrinal factors structure modes of behaviour usually regarded as un- or under-determined: gut instinct and spontaneity. The research discussed in this chapter similarly seeks to identify the determinants of behaviour at the level of the 'genetic,' but it uses this term in the Bourdieusian sense, referring to the internalisation of political and cultural structures, their systematic misrecognition as such and their determination of shared cultures of practice at the non- or pre-conscious level.[3] Specifically, it will assess the extent to which the corporeal and conscious experience of conflict by field actors—in this case, war correspondents—can be explained by latent struggles for authority and power both within the journalistic field and the broader fields of war and cultural production, struggles which take the form of symbolic political economies constituted as professional identity and ideology. Several contributors at the WVW5 conference stressed the need to overcome the tendency in war studies, organisational and Rational Actor Theory (RAT) analysis to view war as a homogenous, static category and to take into account the different historical, political and geographical contexts of conflicts. This chapter extends this proposition into the realm of phenomenology, arguing that not only does time- and space-specific context matter, but so too does the day-to-day temporal aspect of the lived experience of war. In particular, it will discuss the extent to which the durability of certain world-views and perspectives on professional identity rest upon routinisation and whether the unfamiliar experience of boredom and new communication technologies might disrupt existing journalistic cultures of practice.

2. Theoretical Context

Bourdieu's genetic structuralism, like other phenomenological approaches, seeks to supersede the opposition between determinism and voluntarism. It rejects the idea that there is such a thing as teleology of structure, which means that a phenomenon such as political determination or domination has no meaning beyond its enactment in practice—though, since this enactment is effectively necessitated, structures of domination certainly have an endurance which exceeds their discrete moments of constitution.[4] Likewise, it precludes the idea of individuals or groups having 'ownership' of their practices and being able consciously to enact them in ways which avoid complicity in relations of domination and dominatedness. Rather, practices are structured—contextually contingent and shared—but also structuring, the form of which is not linearly determined but a 'particular instance of the possible'[5] and hence, to a finite degree, unpredictable. What this means in practice is that while we cannot simply treat a field's genetic (or generative) structures as an algorithm to calculate resulting practices, we can nonetheless assert that those practices have reason and posit the conditions of possibility which underpin their emergence. In short, we cannot predict individual actions, but nor are we surprised by their observation, each action being an instance of a range of possible behaviour. Bourdieu uses the term *habitus* to capture the shared dispositions which precede and outlast the individual, which are structured according to objective relations of power in a field, which determine the experience of normality, common sense and instinct and which are transposable to situations ranging from the mundane to the counter-intuitive.[6] The upshot is that *habitus* constitutes and is constituted by practice structured according to hard-wired anticipation:[7] the actor either consciously responds or, more commonly and significantly, reacts in a nonconscious, corporeal,[8] practical manner to a lived situation which is experienced as natural but which tends, sometimes haphazardly, to concretise essentially arbitrary norms and symbolic forms associated with contingent hierarchies of power as universal and unremarkable.

The same logic explains the occupation of positions of power (and lack of power) in fields (in this case of war, journalism and cultural production—the co-extensivity of these is discussed below) and the configurations of 'symbolic capital'[9] associated with these positions, which amount to the criteria by which success or value is systematically recognised. It is not the case that any individual or institution could feasibly claim a position of authority as their own; rather, it should be seen as no surprise that the individuals and groups observed in these positions are there. They are also particular instances of the possible: journalists whose socialisation, upbringing, education and so on make their entry to the field of, in this case, war correspondence, a 'natural fit,' and their progression a seemingly natural trajectory. This means that while we should not interpret a particular journalist's success or status as arbitrary, we can regard two journalists in similar field positions as effectively interchangeable.[10] After Weber,[11] the embodiment, personification and projection of recognised valuable forms of symbolic capital—what is perceived as 'good war reporter'—means that success is not seen as the result of the accumulation of skills, but rather an expression of personal character. The well-regarded journalist is then naturally gifted rather than experienced or well-trained; their nose for a story is a gift. Likewise, the *criteria* by which such things are judged are normalised to the point that they are more or less inexpressible—too obvious to bear articulation. However, in this research I raise the possibility that such a nose or 'gut instinct'[12] can be seen as the end product of engagement in the struggle for authority—i.e., a field determination whose particular form is not associated with some uncontested idea of what constitutes good journalism, but rather with a symbolic economy or 'game'[13] whose rules and objectives operate autonomously. This 'game' is experienced as real, needless to say, which in phenomenological terms is the equivalent of saying that the symbolic completes the real and exceeds the material. It is not, to reassert, a wholly arbitrary symbolic world, but one of a number of worlds rendered possible by 'objective' or material context. The next section assesses some of the phenomenological determinants of a journalist's experience and how these may be conceived as stakes within the journalistic field and between journalism and adjacent fields.

The difficulties of putting this political phenomenology into methodological practice are various, but can be grouped into two themes. First, the theoretical account of contingent structures and norms which are internalised and forgotten as contingent commits the researcher to establishing the determinants of categories of experience which inevitably largely defy expression: normalcy, common sense, instinct, trivia and so on.[14] I detail below the methods used to identify and interpret practices whereby individuals invoke tropes which are conceived as universal, uncontroversial—and simply too obvious to say. Second, Bourdieu directs us to believe that the operation of a symbolic economy and its practical universalisation systematically mask ongoing relations of political domination and dominatedness, in which individuals are essentially conceived as 'rational fools.'[15] Towards the end of this chapter I demonstrate how individual reflexivity can be used to add perspective to a model which often seems to suggest that we should treat lack of evidence of political determination—in expressions of the mundane or common-sense—as masking some 'deeper' determination.[16] There is no shortage in the literature, after all, of accounts of professionalism which argue that codes of conduct, professional identities and ideologies emerge not naturally in relation to that profession's object, but to maintain the endurance of elites within the profession and gatekeeping structures to protect it from external competition.[17] In the case of journalism, scholars such as Matheson[18] argue that journalistic ethics are not merely disinterested principles regulating journalists' interactions with their sources, but also the basis on which journalists vie with one another for success, prestige and power. What the methodology of genetic structuralism uniquely allows for, however, is the reconstruction of *how* ethics, professional values and identities are decontested, embodied, naturalised and, potentially, negotiated.

3. Summary of Research Findings

A Bourdieusian analysis of war reporting amounts to establishing the structures and preconditions of the journalistic *habitus*, the adaptable, transposable set of dispositions which determine the individual's behaviour in, and experience of, the theatre of war. This is in part describable in terms of professional and personal values regarding conflict and its media representation, but it also refers to a 'lifeworld'[19] in the sense developed by Husserl and Schutz—that which comprises the taken-for-granted, the sense of normalcy which underpins a journalist's experience of even the most seemingly exceptional circumstances. This includes the senses of corporeality and temporality, as well as what might be termed the limits of linguistic explicability—that is, the threshold of what may be commonly assumed between interlocutors, beyond which lies stating the obvious—as well as taste, personal proclivities and socialisation.[20]

Necessarily these categories are methodologically problematic to access, though also significant, since their embodiment and performance are given in Bourdieusian theory to be politically implicated. The first principle necessary to understand the determination of the correspondent's experience of war is that the phenomenal 'subject' of journalism is not the individual journalist who produces an article or broadcast. Writing about the artistic field, Bourdieu notes that:

> …the 'subject' of the production of the art-work—of its value but also of its meaning—is not the producer who creates the object in its materiality but rather the entire set of agents engaged in the field. Among these are the producers of the works classified as artistic (great or minor, famous or unknown), critics of all persuasions (who themselves are established within the field), collectors, middlemen, curators etc, in short, all those who have ties with art, who live for art and, to varying degrees, from it and who confront each other in struggles where the imposition of not only a world view but also a vision of the artworld is at stake and who through these struggles, participate in the production of the value of the artist and of art.[21]

Accordingly this research, while relatively small-scale, was based on interviews with those who have a 'stake' in determining the cultural value and authority of war reporting as including not only war reporters but also editors, other journalists, military officers, military public relations personnel and government press officers. The analysis began by coding for explicit ascriptions or negations of value (in relation to individuals, institutions and practices), alongside the linguistic style of the speech act: use of metaphor, irony, euphemism, humour, moral judgement, outrage, offhandness and so on.[22] It also catalogued instances of overt versus latent agency -

> So the Americans were firing on them [Iraqi troops] from over there, behind the hill [Interviewee 5]
>
> *versus*
>
> There was an instance where the coverage of the killed and wounded coming back from a friendly fire incident was not allowed and facilitated [Interviewee 7]

- elisions, inclusions and exclusions (especially those implying an us/them distinction) and frames of reference. This allowed for a tentative initial delineation of what would have to hold in order for the observed speech acts to make sense—that is, the unspoken preconditions of the linguistic cultures of practice of the interviewees. The boundaries of this 'discourse' are thus established by inferring what sort of statement would count as nonsensical (the 'division of madness,' in Foucault's terminology[23]) and, more importantly, what is taken for granted. Sometimes this was physically manifest in gestures expressing the speaker's frustration at being unable to put into words that which to them simply *is*. The method followed here assumed that if something is literally too self-evident to put into words, this is not so much a natural limit beyond which explanation turns tautologous as a decontestation of what counts as obvious in a particular context.

More significantly, however, it is implicit *distinctions* made between symbolic forms (values, individuals, institutions and so on) which indicate the ordinariness of the criterion upon which distinction is based.[24] For example, comparing two colleagues on the basis of, say, integrity does not in itself normalise perception that a particular journalist has integrity, but it does normalise integrity as a comparative category. The interviews began by asking respondents to make explicit assessments and comparisons, but were fairly loosely structured to allow the dialogue to move on from conscious explication to more implicit modes of distinction. No interviewer can claim not to constrain an interviewee's speech or to access their ordinary language unproblematically, but it does appear that the more subtle distinctions indicating embodiment of (contingent) indicators of authority occur most frequently when the respondent is not answering direct questions *about* authority.

While this research is primarily concerned with *how* distinctions are performed, it is useful for the sake of illustration to look at particular values which have currency in the field of war reporting. One of the dominant forms of distinction for war reporters in this research was, unsurprisingly, autonomy. But this did not manifest itself in overt descriptions of independence from undue external influence, but rather through a variety of performances distancing the speaker from fellow journalists. This often took the form of identification or perceived professional equality with the military:

> They're [US special forces] like, 'We're professionals, you're professionals,' there's that understanding and they're happy to just you know leave you to it. [Interviewee 2]

> …and they're [US special forces] like, hey he's known, he's a cool guy. Once they know who you are, once they know you're cool they like completely trust you…. [Interviewee 10]

> …and there were American special forces and Brit special forces and they said if you got this far, fine, you know, don't get killed and it worked okay. [Interviewee 5]

Further, several interviewees expressed sympathy with senior military personnel for having to deal with other journalists -

> I do sympathise to some extent because the pillock factor rises amongst journalists and I mean I say that as a 58-year-old who's been doing it for a long time, but I don't think I was as much of a pillock as the ones I see now. I think it might be quite a pain being a PR man, even if you're the greatest PR man, subtle, gentle and understanding, probably eventually the journalistic pillocks piss you off so much that you become unreasonable. [Interviewee 5]

> A lot of journalists are just really stupid. Especially these guys at the BBC, they think they're God's gift to journalism and they piss him [military commander] off and they piss me off too. [Interviewee 2]

—identifying themselves with external authority on the grounds that each sees through the media/war 'game' and 'gets' the contradictions, absurdities and moral ambiguities which characterise any conflict situation.

> You gotta understand that the military is a machine. Once you get to know how it works it all makes sense, it's got its own sort of logic. [Interviewee 10]

Nearly all war reporters interviewed claimed an outsider's perspective on the game in which their peers were enveloped, a game of access and second-guessing which has little to do with 'real' reportage. This perspective was enacted, presumably unconsciously, through a range of linguistic devices including metaphor, metonymy, elision, ironic distance and humour. Pool journalism was widely described using tourism metaphors, for example:

> They were like, 'We'll tell you if anything's going on, if you're especially nice to us we'll give you a trip so you can see something or other.' [Interviewee 2]

> The authorities have organised an excursion to cover the protests.... [Interviewee 6]

This does not indicate open revolt against the pool system, but rather a wry reflexivity which positions the speaker in opposition to it in an unremarkable, naturalised way. Similarly, it was common for reporters to criticise others for being taken in by the symbols of war, whether it be (male) correspondents being overly impressed by military machinery and soldierly camaraderie—

> Anyway he was covering Kosovo and I think he was a journalist who had been tricked into becoming too close to the armed forces and I think it happens a lot with especially male journalists, who like to see themselves as one of the boys. [Interviewee 4]

—or reporters being over-excited at being given camouflage clothing:

> Bottom line was that a lot of the press you know thought this was wonderful, because they were given backpacks and they were you know given chocolate chip trousers and jackets and all the fucking paraphernalia of military status … a lot of journalists love it, they love it, they get sucked in by the military, which is of course the idea. [Interviewee 5]

Here, beyond the explicit dismissal of others' lack of professional distance, there is a more subtle signification in which the speaker does not describe a professional active resistance to constraints on movement and access, but rather a personal, common-sense world-weariness. It can be inferred that substance has less to do with the acquisition and exercise of journalistic skill than perceived personal integrity. This is broadly backed up by distinctions between substance and 'fluff:'

> But TV, fucking fluff monkeys. On TV all they want to do is put on familiar faces, instead of good journos. It's all about face time to them, so they put these fucking fluff monkeys out there. See with TV the standards are lower and that's producers too, not just the journos. There's not as many of 'em and they don't know anything. It's like you'll get a guy and it's like the most exciting thing he'd ever done was the floods in Chicago or something. [Interviewee 10]

> You know, some of 'em are more used to human interest stories and that shit, you know, 'cause that's what they're used to, it's all nice stuff about Private Johnny and his kids back home in Nebraska. [Interviewee 10]

> …that's not a press issue, that's not the kind of thing that I think about, it's analysis, it's soundbites on television. What my concern is, what news editors' concerns are, reporters' concerns are news reporting, access to news, not whether someone calls this a quagmire on the editorial page. That is true, that is carried in newspapers and in all 24 hour cable channels, that is not news, that is thumb sucking. [Interviewee 3]

The analysis concluded that seeing through the journalistic game was never a simple statement of fact, but always a positioning—a game in itself, which proceeds according to two symbolic economies. The first is an economy of esotericisation, by which journalistic authority consists neither in affirmation of or opposition to specific values, but something which just is:

> He's just a better journalist; he always had a feel [pause, gestures] for the job which Fialka doesn't. [Interviewee 1]

This is not an original insight—it is in line with the theory of professionalisation proposed by Johnson[25]—but it does demonstrate the mechanics of how esotericisation proceeds as well as its politicality, not only in the traditional sense of gatekeeping but also in the decontestation and embodiment of specific ethical commitments whose dominance in journalistic discourse is not natural. The second economy is one of ambivalence towards power, money and, to a degree, morality (or at least moralising). Authority in this economy consists in irony and cynicism, a distance which is more of a personal positioning 'above' debates over ethics and professional values than a positive defence of objectivity. This frequently takes the form of offhandness in descriptions of dangerous situations or 'serious' issues:

> ...journalists who have been wounded or killed or otherwise prevented from doing their job. [Interviewee 8]
>
> They pointed their guns at us and shouted, 'Who are you? You don't have credentials.' And we were like, 'This is Pristina, we don't need credentials.' But it was all quite fun though. [Interviewee 13]
>
> ...you could get shot at and strafed if you played your cards right. [Interviewee 5]
>
> Drew would be great if you want to talk to the bang bang guys. [Interviewee 10]

This was often observable in characterisations of socialisation, or rather a personal aversion to social interaction and statements along the lines of 'does not play well with others.' The naturalised performance of an innate misanthropy appears significantly related to the perception of journalistic authority and would be compatible with traditional romanticised myths about the war correspondent such as rugged individual or loner. The contingencies upon which such embodiment is based (e.g., demeanour, language) are given to be systematically misrecognised, as is the essential arbitrariness of the connection between the symbolic form of that identity and its relative position of power in the field of war correspondence.

However, there are three factors working against this conclusion, each of which may potentially lead us to a more nuanced account of the correspondent's experience of war and away from a more deterministic model in which all aspects of that experience, including their everyday consciousness, their sense of time and space, instinct, spontaneity and the sense they make of conflict, are structurally determined and politically suspect. First, there was evidence in the interviews of reflexivity whose content exceeded its strategic function. It is always possible to interpret speech acts as primarily strategically oriented, but it is also important to defend the principle that not all observed behaviour is meaningful in the narrow sense of pre-defined categories of interpretation. Thus, it is feasible to characterise comments such as the following as another example of a journalist engaging in a seeing-through-the-game practice which is collectively valued in the journalistic field (and the broader field of cultural production):

> …now everyone thinks the way you present things is you know with a furrowed brow and an earnest manner….[Interviewee 5]

> …these guys who come back from Afghanistan with their scarves around their necks, a little bit of dust on their shoulders…. [Interviewee 14]

However, it is impossible to rule out the possibility of any substantive insight without entering into a circular debate over strategic practices in journalism and academia. It might be difficult to disaggregate the different functions of this sort of comment, but it serves little purpose to insist that it has nothing significant to say *about* war reporting.

Second, several critics have argued that Bourdieusian method effectively assumes that a professional field is hermetically sealed, or that we can conceive of journalistic culture as existing in a vacuum.[26] In fact, Bourdieu[27] claimed that journalism is only weakly autonomous within the field of cultural production, meaning that while it has some sort of internal logic which determines valorised symbolic forms and structural reproduction, it is more exposed than most cultural fields to external forces—especially market forces. Further, it need not be assumed that journalists inhabit single fields rather than being simultaneously immersed in several. Journalism by definition involves interaction with sources from other fields and this research was designed upon the premise that in order to understand the symbolic value of a cultural form one must take into account, in Bourdieu's words, all those who have a stake in its determination. This enriches rather than limits the scope for analysis, leading to the exploration of tensions and commonalities of symbolic economy between worlds which co-exist and vie for power. I would argue that there is another productive line of inquiry opened up by this multi-field approach: the multiplicity of roles within the individual's experience of war. This means seeing the journalist not only as a cultural producer but also as a consumer and social actor negotiating relations and tensions between public, professional and private spheres. These relations are changing according to political and cultural shifts (the decline of the separation between the public and private sphere, the rise of individualism) and the radically disrupted and transformed modes of communication, networks, individual discreteness and separation ushered in by new technologies. I return to these themes below.

Third and perhaps most germane, war is not static. In order to ascertain what determines an individual's conscious and nonconscious experience of a given conflict, it is important that the methodology does not assume that conflicts are uniform. The interviews were initially designed to ask reporters specifically about the 1990-91 Gulf War,[28] but it soon became apparent that wars did not exist as discrete units in their memory, instead they form part of a narrative which is part individual and part collective. It is this narrative which reveals most about how the phenomenology of war correspondence has changed since the most senior respondents entered the field in the 1960s. In spatial terms and staying with the Gulf War for the time being, what is interesting is how the prevalence of the pool system (used previously in Granada, the Falklands and Panama but for different reasons and to a lesser extent) influences dominant modes of distinction. 'Independents' were lionised as in other conflicts, but, as noted above, there was also scope for distinction in the absence of mobility in the form of cynicism, irony and so on. Likewise, identification with military personnel or the performed reflexivity about relations with soldiers is contingent upon a spatial environment of close physical proximity. This does not preclude the idea of professional distance, but it does change the practices through which it is manifested. In short, distinction between journalists can be effected both through identification with military personnel (implicitly in contrast to peers) and through maintaining a professional distance from military figures, but this distance is of a stylistic form distinct from previous wars.

4. Space, Time, Subjectivity and Journalistic Autonomy

According to the Bourdieusian model there is no qualitative difference between distinction criteria—their function as field effects are primary—and as such a personal distance projected as disposition is no less effective a structural reproductive mechanism than physical autonomy. But this is not to say that when journalists are deprived of physical autonomy they will simply find other ways to be professional—only that they will find other ways to compete for the symbolic capital of prestige or integrity. In fact, autonomy is undermined by several factors beyond physical restrictions. This in part relates to the feelings of camaraderie, romanticisation of soldiers, or alternatively a form of Stockholm syndrome, discussed at length elsewhere and all cited in the interviews, usually by way of establishing the merit of remaining impervious to such psychologies and being aware that they in fact form part of the military's media management strategy. It also relates to the sharing of symbolic worlds, common interpretative frameworks producing at least a limited intersubjectivity between military and journalism, as much as their identities and professional roles are kept distinct. A common intersubjectivity (that is, a collective, socially determined consciousness) literally precludes the possibility of maintaining a distinct individual subjectivity, since the resources enacted in processes of subjectification are themselves not distinct.

There is not space here to discuss in detail the phenomenology by which the journalistic consciousness emerges out of material and symbolic context, but one key factor is time. Specifically, the stabilisation of any symbolic forms—professional norms, identities and so on—depends on their patterned enactment and habituation over an extended period. In terms of the Gulf War the temporal factor is important for several reasons. First, it imposes a highly structured schedule on journalists whose professionalism is traditionally tied to adaptability, along with a personally relaxed attitude to time. As with the spatial aspect this disrupts one means of projecting professional identity. But it also introduces a regularised alternative which is potentially conducive to a limited ritualisation: that is, the routines of military life may lead to a heavy structuring not only of professional practice but also, inevitably, the performance of identity—a structuring which would be difficult consciously to resist given that it is in part constitutive of conscious experience. This is not to suggest that conflicts are normally characterised by regimental time schedules and hermetic spatial conditions, but it does raise the possibility of comparing conflicts according to their relative 'rationalisation' of the journalistic experience, as Foucault would have it—whether their operation is temporally chaotic or ordered and the impact this will have on the means by which correspondents make sense of events, their surroundings and themselves as journalists. The data on the Gulf War suggests two different trends. For pool journalists, the imposition of a daily routine was vocally resisted, but it does seem to have effected a structuration of the perceived role of the journalist, which is best described in terms of the 'news discourse' which delimits the possibilities—what is conceivable, sayable and doable—of professional practice and identity. This was manifest in a broad lowering of expectations (again accompanied with varying distancing strategies for expressing such), a preconception of what journalists were there to do, which was (at least after the fact) effectively normalised as frustrating but uncontroversial. For embedded journalists, on the other hand, it seems that the key factor in terms of time was boredom—large stretches of time lacking in structure and yet, given the preclusion of alternatives, not unstructured. Bård Mæland's chapter in this book develops a fuller account of the phenomenon of boredom in the military. The conclusions in this regard drawn from the war correspondent interviews are at best tentative, but it does appear from the two reporters who raised boredom that journalists find themselves ill-equipped to deal effectively with it and it seems common for embeds to rely on soldiers' time-killing tactics to cope—potentially, but not definitively, undermining autonomy through simultaneous enactment of cultures of boredom-fighting practice. It would be churlish to suggest that watching DVDs with soldiers undermines a journalist's independence;[29] it may be significant, on the other hand, that the temporal disorientation and disruption embedded journalists experience leads to a convergence of practices. Finally, this is not intended to suggest a normative injunction that any disruption or imposition is bad for journalism—it remains the case that existing cultures, practices, norms and routines of journalism serve to reproduce the hierarchies and power relations of the journalistic field, as much as the field's internal logics may also guarantee the survival of defensible professional values.

5. New Technologies and the Phenomenology of War Reporting

Alongside embedded and pool journalism, perhaps the most significant (and obvious) shift in the factors determining a correspondent's experience of war is technological. The impact of mobile communications technology has clearly changed the way that war reporting is done. In our present context, what is particularly interesting is how new devices and formats have been integrated into existing processes of distinction, socialisation and professional identity. The interviews suggest two countervailing trends. On the one hand, some respondents see new technologies simply in terms of autonomy—particularly that they mitigate the need to rely on the military for transmission of reports. In this regard technology is merely a new resource in an unchanging economy of ingenuity and integrity, summed up well by one interviewee who recalled evading US censorship first by hiring a satellite telephone in northern Iraq and then, on finding it more of a hindrance than a benefit (this was in 1991 and the phone was 'enormous'), negotiating with Jalal Talibani to use the Kurds' telegraph service—the specific technology used is secondary to the use of guile to elude restrictions on journalistic practice. For others, particularly more senior interviewees, technology becomes an explicit component of individuation (and by association individual professionalism) in terms of its dismissal. Several interviewees saw new communication technologies through the aforementioned prism of substance/fluff, characterising devices as 'gadgets,' 'gizmos' and 'toys' and their failure to fall for them as indicative of personal straightforwardness which is immune to the vicissitudes of fashion. I would suggest that this scepticism is in line with the economy of ambivalence set out above, a performance of a dominant journalistic disposition which is not easily impressed and quick to see through marketing tactics and ulterior motives.

But beyond these accounts of how technology relates to professional identity, new communication technologies have the potential to disrupt the whole symbolic economy of war reporting, of which practices of individuation are one subset. Perhaps most significant is the ability to access different sources from those available on the ground, leading to a renegotiation of the importance of spatial proximity and personal immediacy, encapsulated well by the initial reporting of the war in Afghanistan, in which there was a tacit consensus amongst all major networks having a correspondent inside (northern) Afghanistan was an indispensable indicator of authority, notwithstanding that these correspondents were receiving their information not from sources encountered in Afghanistan but from the news desks of London and New York. It can be argued[30] that personal access to sources has attained the status of a mythic good in journalism: there is no need to assume that immediacy of interaction has natural value and indeed phenomenology effectively problematises the notion of immediate experience as well as its prioritisation over mediated experience. This was confirmed by one (relatively young) correspondent who invoked the spectatorship of war reporting, dovetailing with significant recent research[31] into the spectacle of war and suffering both in terms of journalistic 'bearing witness' and media audiences:

> I mean, war is the best spectator sport in the world. It's fun, I love it, it's just such a rush … it's kind of like watching sport, but much more intense. So I kind of tend to say if people ask me why I'm a war correspondent, it's like, well if I didn't I'd have to have a fucking job. I mean, this isn't work, it's fun. I'm getting paid to do all this stuff which is totally amazing and I love it, we all do. [Interviewee 10]

In any case, it does not hold that the proliferation of sources is a good thing: it leads as easily to homogenisation as diversification and in Foucaultian terms can be argued to establish a more restrictive discourse around a conflict through the increased abundance of instruments of discursive reproduction. Most importantly, it is wide of the mark to suggest that the correspondent's experience of war has been transformed from a material to a virtual one. It is certainly true that an increasing proportion of the information gathered by a war reporter comes in mediated form—printed statement, military monitors—but the doing of war reporting remains a material practice. It is tempting to argue that space (and especially distance) will have little to do with journalism in the future, but what is really happening is a re- rather than de-spatialisation. New technologies have undoubtedly disrupted what Heidegger refers to as 'thrownness,' the sensation of being enveloped in the always-already existence of one's surroundings. But this is not a simple opening up of the determination of the journalistic consciousness or a freeing of the journalistic mind from the substantial constraints of conflict situations. Instead, it should be seen as a different determination—it is not a question of lionising immediacy or multiplicity or fluidity in journalistic practice, but rather of understanding how new modes of communication alternately structure the conscious experience of the journalist and what consequences this might have. The analysis suggests four possibilities.

The first regards the internal power hierarchies of the journalistic field, in particular between journalists and editors. The authority of war (and to an extent foreign) correspondents is partly tied to the confidence placed in them to practice journalism in highly unpredictable situations without much in the way of supervision or management. Correspondents interviewed did not report that their editors were calling constantly to check up on them, but there has been a palpable shift in what is regarded as unremarkably expected in two ways: the frequency with which a journalist is expected if not to file a story then at least to provide detailed information and the level of detail expected regarding forward planning. This does not amount to an editorial coup, but might suggest a subtle change in which war correspondents, through being 'incited' (in Foucault's words) to communicate, are rationalised and managed. As with other constraints detailed above, there is room in this disposition for performances of distinction, evident from one interviewee comparing his editor to worried parents.

Second, it is possible that the increasing accessibility and visibility of the war correspondent means that it is no longer feasible for the experience of conflict to be systematically mystified. War reporters still often make the claim that others cannot understand what it is like to 'be there,' no matter how explicit conflicts have become. There is evidence that fellow journalists no longer regard war reporting as a mystified or esoteric realm, as was prevalent before, but causation is impossible to demonstrate: it could be the product of increased visibility, but it is equally possible that it is the result of a changing mainstream journalistic culture in which the norms of war reporting are seen as archaic or anachronistic.

A third trend and a significant one in terms of the consciousness of the war correspondent, is that practices of journalistic production are increasingly interspersed with other practices—especially consumption. It is more and more common for journalists to read news and other websites in the field and—not dissimilar to office workers—to do so at regular intervals across a working 'day.' This temporal embedding suggests a different mindset, one more grounded in 'home' news discourses, narratives and cultures, potentially with a more naturalised, routine awareness of peer and audience expectations. While I have suggested here that the collapse of distance should not be exaggerated, such an overlapping or synchronisation with 'domestic' cultures of news production and consumption would be expected to impact on the subjective experience of the war correspondent, partly determined as it is as an intersubjectivity common to news audiences, as well as those of journalistic producers and (as we have seen here) the military. In research terms this could form a productive link to audience studies investigating what viewers and readers are *doing* when they consume war reporting.[32] It also points to recent research into the blurring of the boundaries of production in war journalism brought about by the increasing prevalence of media content produced by soldiers and distributed online.[33]

Finally and relatedly, it was noted above that Bourdieu is often accused of overstating the significance of professional fields, and underplaying the fact that individuals are simultaneously immersed in several fields professional and otherwise. The advent of mobile communications technology offers the opportunity to correct this, providing evidence that, like the practices of consumption mentioned above, journalists serve a multitude of roles in non-professional practices that are intertwined with the journalistic life—especially communicating with friends and family. There was no shortage of evidence in the interviews that the war correspondent disposition—which I have argued here is to be interpreted as a set of strategic practices rather than an identity *per se*—is one singularly unsuited to domestic and family life. This could be explained in terms of the moral ambiguity which may signify 'natural' authority in the field, or in terms of the 'lone wolf' archetype in which war reporting's cultural cachet is traditionally invested, but insofar as it is experienced not as strategy but as personally, corporeally, immediately real, communications technology would be expected to impact significantly on the journalistic consciousness insofar as it is constituted socially, through interaction with others. Perhaps more saliently, it demonstrates the increasing interconnectedness not just of journalism and other professional and cultural fields, but of the different aspects of individual subjectivity—public and private selves—which is characteristic of broader journalistic trends towards casual and freelance work practices and the proliferation of non-professional media production.

6. Conclusion

Several papers presented at the conference stressed the need to construct an inter-disciplinary approach to war studies, in Harmon's words drawing on 'fields as diverse as biology, ethology, game theory, behavioural endocrinology, cognitive neuroscience, memetics, psychology, choice theory [and] anthropology' to supplement standard political analyses of conflict. Further, Nick Mansfield and Jason Edwards argued in their chapters that 'political' analysis itself needs to be disaggregated, moving from historically and economically determinist studies of policy-making and zero-sum approaches to conflict strategy to encompass political communication, network theory and postmodernist and poststructuralist problematisations of the political. This is very much in line with the broader move in political studies towards a supersession of the artificial opposition between qualitative and quantitative perspectives, as well as that between continental and analytical philosophies. It also, with some irony, brings us back to Clausewitz's dictum that war should not be seen as *sui generis* but rather as continuous with society and politics—and as such subject to the full range of social and political analysis. The political phenomenological approach taken here is intended to be a contribution to this inter-disciplinarity. It aims to explain the conscious experience of war, in this case by journalists, in terms which neither regard consciousness as deontological nor reduce it to brute mechanics. Instead, it sees the experience of war as determined by a broad range of factors—some predictable, some chaotic and usually political in the sense that all individuals and institutions active in the field of war have a stake in how it proceeds. In this chapter it has been seen that changes in how individuals inhabit their environment—such as physical constraint and temporal deroutinisation—disrupt the normal unspoken rules by which they interact and produce distinctions. This only serves to emphasise that the rules of intersubjective engagement between war correspondents, other media professionals, military personnel and political actors are far from uniform and are determined variously by historical context, geography and increasingly the technological infrastructure underpinning communicative interaction.

Notes

[1] Nick Mansfield explored the limitations of this tradition in his conference paper 'Fighting for Peace: From the Social World to Armed Democracy', *The Metamorphosis of War*, A. Plaw (ed), Rodopi, Amsterdam and New York, NY, 2012, p. 4.

[2] H. Tumber & F. Webster, *Journalists under Fire: Information War and Journalistic Practice,* Sage, London, 2006.

[3] For Durkheim, the social must be explained not by the conceptions of its participants, but by the structural causes which elude awareness but which necessitate the phenomena observed by the social scientist. For Bourdieu, all observed behaviour (including participants' conscious reflections) should be seen after Bachelard as specific expressions of common generative structures which are the key objects of social analysis.

[4] Some critics, however, argue that Bourdieu fails to overcome a deterministic teleology of political reproduction. See, for instance, J. Bohman, 'Reflexivity, Agency and Constraint: The Paradoxes of Bourdieu's Sociology of Knowledge,' *Social Epistemology*, Vol. 11, No. 2, pp. 171-186.

[5] P. Bourdieu, *Practical Reason: On the Theory of Action*, Polity Press, Cambridge, 1998, p. 2.

[6] P. Bourdieu, *Outline of a Theory of Practice*, Cambridge University Press, Cambridge, 1977, p. 72.

[7] 'The practical world that is constituted in the relationship with the *habitus*, acting as a system of cognitive and motivating structures, is a world of already realised ends…' P. Bourdieu, *The Logic of Practice*, Polity Press, Cambridge, 1990, p. 53.
[8] In Merleau-Ponty's words, 'We said earlier that it is the body which understands in the acquisition of habit. This way of putting it will appear absurd, if understanding is subsuming a sense datum under an idea and if the body is as an object. But the phenomenon of habit is just what prompts us to revise our notion of 'understand' and our notion of the body. To understand is to experience the harmony between what we aim at and what is given, between the intention and the performance—and the body is our anchorage in a world.' M. Merleau-Ponty, *Phenomenology of Perception*, Routledge and Kegan Paul, London, 1962, p. 167.
[9] P. Bourdieu, *The Logic of Practice*, op. cit., pp. 112-121; P. Bourdieu, *Language and Symbolic Power*, Basil Blackwell, Oxford, 1991, pp. 72-76.
[10] P. Bourdieu, *Outline of a Theory of Practice*, op. cit., pp. 85-86.
[11] M. Weber, *Economy and Society: An Outline of Interpretative Sociology*, Bedminster Press, New York, 1968.
[12] I. Schultz, 'The Journalistic Gut Feeling,' *Journalism Practice*, Vol. 1, No. 2, 2007, pp. 190-207.
[13] P. Bourdieu, *The Logic of Practice*, op. cit., p. 53.
[14] See, for instance, H. Kögler, 'Alienation as Epistemological Source: Reflexivity and Social Background after Mannheim and Bourdieu,' *Social Epistemology*, Vol. 11, No. 2, 1997, pp. 141-164.
[15] A.K. Sen, 'Rational Fools: A Critique of the Behavioural Foundations of Economic Theory,' *Public Affairs*, Vol. 6, No. 4, 1977, pp. 317-344.
[16] T. Markham and N. Couldry, 'Tracking the Reflexivity of the (Dis)Engaged Citizen: Some Methodological Reflections,' *Qualitative Inquiry*, Vol. 13, No. 5, 2007, pp. 675-695.
[17] N. Carpentier, 'Identity, Contingency, Rigidity: The (Counter)Hegemonic Constructions of the Identity of the Media Professional,' *Journalism*, Vol. 6, No. 2, pp. 2005, 199-219; M. Deuze, 'What is Journalism?' *Journalism*, Vol. 6, No. 4, 2005, pp. 442-464; H. Gans, 'Everyday News, Newsworkers and Professional Journalism,' *Political Communication*, Vol. 24, No. 2, 2007, pp. 161-166; H. Pöttker, 'Objectivity as (Self-)Censorship: Against the Dogmatisation of Professional Ethics in Jouralism,' *Javnost: The Public*, Vol. 11, No. 2, 2004, pp. 83-94.
[18] D. Matheson, 'Scowling at Their Notebooks: How British Journalists Understand their Writing,' *Journalism*, Vol. 4, No. 2, 2003, pp. 165-183.
[19] 'Lifeworld' (*Lebenswelt*) is used in the Husserlian sense of a taken-for-granted stream of everyday routines, interactions and events that constitute

individual and social experience, rather than Habermas's conception of shared understandings and values which develop over time through face-to-face communication in a social group.

[20] Goffman writes: 'To walk, to cross a road, to utter a complete sentence, to wear long pants, to tie one's shoes, to add a column of figures—all these routines that allow the individual unthinking, competent performance were attained through an acquisition process whose early stages were negotiated in cold sweat.' E. Goffman, *Interaction Ritual: Essays on Face-to-Face Behaviour,* Allen Lane, London, 1972, p. 293.

[21] P. Bourdieu, *The Field of Cultural Production: Essays on Art and Literature*, Polity Press, Cambridge, 1993, p. 261.

[22] P. Bourdieu, *Language and Symbolic Power*, op. cit., pp. 58-96.

[23] M. Foucault, *The History of Sexuality, Vol. 1*, Penguin, Harmondsworth, 1981, p. 55.

[24] This derives from Saussure's characterisation of language as a system of differences and meaning as being constituted by difference.

[25] T. Johnson, *Professions and Power*, Macmillan, London, 1972.

[26] R. Benson, 'News Media as a 'Journalistic Field': What Bourdieu Adds to New Institutionalism and Vice Versa,' *Political Communication*, Vol. 23, No. 2, 2006, pp. 187-202.

[27] P. Bourdieu, 'The Political Field, The Social Science Field and the Journalistic Field' in *Bourdieu and the Journalistic Field*, R. Benson & E. Neveu (eds), Polity Press, Cambridge, 2005, p. 33.

[28] The correct name for this conflict is disputed. This paper follows the convention that the 1980-88 conflict is referred to as the Iran-Iraq War, 1990-91 as the Gulf War and 2003 onwards as the Iraq War.

[29] See also E. Coble's conference paper abstract, 'Entertainment and Understanding: The Evolving Role of Movies in the Combat Zone.'

[30] See, for instance, H. Pöttker, op. cit.

[31] See especially M. Aaron, *Spectatorship: The Power of Looking On*, Wallflower Press, London, 2007; L. Chouliaraki, *The Spectatorship of Suffering*, Sage, London, 2006.

[32] L. Boltanski, *Distant Suffering: Morality, Media and Politics*, Cambridge University Press, Cambridge, 1999; P. Hammond, *Media, War and Postmodernity*, London, Routledge, 2007.

[33] C Christiansen, 'Uploading Dissonance: YouTube and the US Occupation of Iraq,' *Media, War & Conflict*, Vol. 1, No. 2, 2008, pp. 155-175.

Bibliography

Aaron, M., *Spectatorship: The Power of Looking On*. Wallflower, London, 2007.

Benson, R., 'News Media as a 'Journalistic Field': What Bourdieu Adds to New Institutionalism and Vice Versa.' *Political Communication*. Vol. 23, No. 2, 2006, pp. 187-202.

Bohman, J., 'Reflexivity, Agency and Constraint: The Paradoxes of Bourdieu's Sociology of Knowledge.' *Social Epistemology*. Vol. 11, No. 2, 1997, pp. 171-186.

Boltanski, L., *Distant Suffering: Morality, Media and Politics*. Cambridge University Press, Cambridge, 1999.

Bourdieu, P., *Outline of a Theory of Practice*. Cambridge University Press, Cambridge, 1977.

—, *The Logic of Practice*. Polity Press, Cambridge, 1990.

—, *Language and Symbolic Power*. Basil Blackwell, Oxford, 1991.

—, *The Field of Cultural Production: Essays on Art and Literature*. Polity Press, Cambridge, 1993.

—, *Practical Reason: On the Theory of Action*. Polity Press, Cambridge, 1998.

—, 'The Political Field, the Social Science Field and the Journalistic Field.' *Bourdieu and the Journalistic Field*. Benson, R. & Neveu, E. (eds), Polity Press, Cambridge, 2005.

Carpentier, N., 'Identity, Contingency and Rigidity: The (Counter-) Hegemonic Constructions of the Identity of the Media Professional.' *Journalism*. Vol. 6, No. 2, 2005, pp. 199-219.

Chouliaraki, L., *The Spectatorship of Suffering*. Sage, London, 2006.

Christiansen, C., 'Uploading Dissonance: YouTube and the US Occupation of Iraq.' *Media, War & Conflict*. Vol. 1, No. 2, 2008, pp. 155-175.

Deuze, M., 'What is Journalism?' *Journalism*. Vol. 6, No. 4, 2005, pp. 442-464.

Foucault, M., *The History of Sexuality, Vol. 1*. Penguin, Harmondsworth, 1981.

Gans, H. 'Everyday News, Newsworkers and Professional Journalism.' *Political Communication*. Vol. 24, No. 22, 2007, pp. 161-166.

Goffman, E., *Interaction Ritual: Essays on Face-to-Face Behaviour*. Allen Lane, London, 1972.

Hammond, P., *Media, War and Postmodernity*. Routledge, London, 2007.

Kögler, H., 'Alienation as Epistemological Source: Reflexivity and Social Background after Mannheim and Bourdieu.' *Social Epistemology*. Vol. 11, No. 2, 1997, pp. 141-164.

Mansfield, N., 'Fighting for Peace: From the Social War to Armed Democracy'. *The Metamorphosis of War*. Plaw, A. (ed), Rodopi, Amsterdam and New York, NY, 2012, pp. 3-20.

Markham, T. & Couldry, N., 'Tracking the Reflexivity of the (Dis)Engaged Citizen: Some Methodological Reflections.' *Qualitative Inquiry*. Vol. 13, No. 5, 2007, pp. 675-695.

Matheson, D., 'Scowling at their Notebooks: How British Journalists Understand their Writing.' *Journalism*. Vol. 4, No. 2, 2003, pp. 165-183.

Merleau-Ponty, M., *Phenomenology of Perception*. Routledge and Kegan Paul, London, 1962.

Pöttker, H., 'Objectivity as (Self-)Censorship: Against the Dogmatisation of Professional Ethics in Journalism.' *Javnost: The Public*. Vol. 11, No. 2, 2004, pp. 83-94.

Schultz, I., 'The Journalistic Gut Feeling.' *Journalism Practice*. Vol. 1, No. 2, 2007, pp. 190-207.

Sen, A.K., 'Rational Fools: A Critique of the Behavioural Foundations of Economic Theory.' *Public Affairs*.Vol. 6, No. 4, 1977, pp. 317-344.

Tumber, H. and Webster, F., *Journalists under Fire: Information War and Journalistic Practices*. Sage, London, 2006.

Weber, M., *Economy and Society: An Outline of Interpretive Sociology*. Bedminster Press, New York, 1968.

Appendix

No	Age	Gender	Years in journalism	Nationality	Residence	Current Position	Former Position
1	65	Male	N/A	US	Washington DC, US	Think tank senior advisor	US Army general
2	Late 50s	Male	≈30 yrs	UK	London, UK	Columnist, UK broadsheet	War correspondent, broadsheets and tabloids
3	50s	Male	N/A	US	Washington DC, US	Think tank president	Political scientist
4	38	Female	≈15 yrs	Australian	Oxford, UK	Broadcast Journalist, BBC	Broadcast journalist, Australia
5	58	Male	≈35 yrs	UK	London, UK	Broadcast journalist, Non-BBC	War correspondent, broadcast and broadsheet
6	50s	Male	≈25 yrs	US	Washington DC, US	Senior editor (quality newspaper)	International editor (quality newspaper)
7	Late 40s	Female	N/A	US	Washington DC, US	State Department press officer	Government advisor
8	40s	Female	≈20 yrs	US	London, UK	Broadcast journalist, non-BBC	War correspondent, US Television
9	50s	Male	≈20 yrs	US	Baltimore, US	Senior academic	Broadcast journalist/ editor, NPR
10	28	Male	3 yrs	US	Washington DC, US	War reporter, US news magazine	N/A

11	63	Male	≈40 yrs	UK	Washington DC, US	Journalist, quality US newspaper	Journalist, various US and UK broad-sheets/tabloids
12	62	Male	≈40 yrs	UK	London, UK	Senior editor, UK broadsheet	War correspondent, international editor.
13	Late 30s	Male	10 yrs	UK	London, UK	War correspond., UK broadsheets	War correspondent, tabloid
14	50s	Male	≈30 yrs	UK	Washington DC, US	Columnist, UK broadsheet	Journalist/war correspondent, UK tabloids/broad-sheets

Tim Markham is Lecturer in Media (Journalism) at Birkbeck, University of London. His current research focuses on moral authority in the news media and he is co-author of *Media Consumption and Public Connection: Beyond the Presumption of Attention*, published by Palgrave Macmillan in 2007.

Part IV:

Waging the New Wars

'Like a Phoenix from the Ashes': The IRA as a Multi-Generational Movement and Its Relevance for the War on Terror

Timothy D. Hoyt

Abstract
The 21st century and the War on Terror have revitalized interest in terrorist groups, irregular warfare, transnational movements and insurgency. Many analysts and scholars have focused their attention on how terrorism can be defeated and how and when terrorist groups cease to exist. This chapter will examine a different phenomenon—how terrorist or insurgent groups can persist and become multi-generational revolutionary movements. The Irish Republican movement and its armed wing, the Irish Republican Army (IRA), continue to exist after at least four separate armed campaigns, two of which can only be termed abject failures. Surviving both defeat and sometimes partial success, the IRA continued to re-emerge each generation, like a phoenix from the ashes—adjusting its new campaigns each time to an internal analysis of the reasons for its success or failure. Assessing the IRA as both a strategic actor and as a multi-generational movement, therefore, may provide useful insights for how to combat Al Qaeda—another transnational movement based on a seductive but rigid and militaristic ideology—and perhaps how to avoid making a multi-generational movement out of our current adversaries.

Key Words: Terrorism, Ireland, IRA, Sinn Fein, irregular warfare, insurgency, War on Terror, strategy.

*****

1. Introduction

When do terrorist groups win? After the attacks of 9/11/2001, terrorism commentators swarmed from every corner, offering profound and sometimes deeply flawed assessments. Some argued that terror was *never* effective, whether waged by a state or a non-state actor.[1] Others argued that, in the modern era, terrorism had the potential to be much more effective, due to the advancements in communications associated with globalization, information technology and international travel. The issue of how and when terrorists can be successful remains a significantly misunderstood subject.

This chapter will use a comparative case study—the experience of the Irish Republican Army (IRA)—applying the theories of Carl von Clausewitz's *On War* and, to a lesser extent, Sun Tzu's *The Art of War*, to irregular warfare and most specifically to Ireland.[2] Finally, it will attempt to link the IRA case to the war on terror. The IRA represents one of the few cases of multigenerational terrorist groups extant in the literature and may illuminate important issues for policymakers who define the current struggle against Islamic extremism as a 'long war.'[3]

2. Strategy and Irregular Warfare: Some Basic Ideas

On War doesn't mention Ireland at all and pays relatively little attention to irregular warfare.[4] Not surprisingly, Sun Tzu appears to be even less familiar with the Irish problem. Nevertheless, the theoretical frameworks Clausewitz and Sun Tzu use to understand strategy and war are applicable to the actions of terrorist groups and non-state actors. What, after all, are conflicts involving these types of groups if not wars of the weak—efforts by a weaker side to impose their will on a stronger adversary by force?[5] Terrorism, insurgency and irregular warfare are all *means*—types of violence applied in an effort to achieve some political objective. As will be discussed below, the IRA's political objective remains constant, based on a nearly 150-year old tradition of violent struggle.

Creating a successful strategy to win against a stronger adversary requires two key elements that are stressed not only in *On War* but also in Sun Tzu's ancient strategic text *The Art of War*—rationality and interaction.[6] Rationality must be considered not in a psychological sense—where terrorism is involved, utilizing the deliberate massacre of innocent civilians, psychological rationality is particularly problematic—but from a more practical perspective: the relationship between ends and means, the managing of costs and benefits and the consideration of risks and potential rewards. A rational terrorist or insurgent (again, recognizing the potential cognitive dissonance in that term) will carefully weigh the consequences of violent action and the potentially counter-productive effects that certain types of violence may have on the possibility of achieving the intended political aims.

The second element of strategy is understanding interaction—the fact that war is not waged against a potted plant, but instead against an active and intelligent opponent.[7] This is particularly vital for smaller states and non-state actors. Their opponents begin the struggle with vastly greater resources which, if marshalled effectively, could crush the insurgent early in the conflict. The weaker combatant, therefore, must fight intelligently, always seeking the opportunity to capitalize on the mistakes of the adversary. Weaker sides may also fight defensively—Clausewitz stresses that the defence is the stronger form of war—and attempt to protract the conflict in hopes that the situation will change and become more favourable.[8] This does not imply passivity—indeed, Clausewitz stresses that the objective of defensive tactics is to eventually transition to the offensive.[9] The idea is to force the enemy to make mistakes and to capitalize on the adversary's errors, in ways that increase the weaker side's options and reduce the overall advantage of the stronger side. Terrorism can be a useful part of this strategy in some circumstances, forcing an enemy to overreact in a way that harms the stronger side's legitimacy and authority or, by persistent attacks, demonstrating the ineffectiveness of existing political structures.

The principles of interaction and rationality, therefore, must guide the combatant—but mean relatively little in the absence of more precise calculations—something Sun Tzu emphasizes almost to the point of exaggeration.[10] Clausewitz states specifically that the first and most important thing the leader and commander must do before undertaking a war is to understand its nature.[11] The tools he offers to measure this understanding are particularly valuable for the study of terrorist and insurgent success.

3. Terror, Insurgency and the Clausewitzian 'Paradoxical Trinity'

Clausewitz describes the nature of a given war—as opposed to the nature of war proper, a much more complex topic—in terms of three major forces: rationality, often associated with the government or political leadership; the forces of chance and probability, often associated with the realm of military struggle; and passion and hatred, often associated with the populace.[12] A state preparing to go to war can estimate the likelihood of success by examining these elements within its own political society. What are the chances of military success? Can it win quickly and decisively? If not, does it have the passion and public support for a long war? Defensive wars may manufacture this support more easily than offensive wars—people may be more willing to pay a higher cost (measured in magnitude of effort and duration) to defend what they have or recover what has been taken from them than to take something from someone else.

Finally, rationality is the art of exerting political control over the war, of understanding when means and ends are misaligned and of adjusting either or both to achieve a favourable conclusion to the conflict. Hatred may inspire a people to fight in ways that increase the chance of catastrophic loss or that violate the moral and ethical principles on which a society is based (see Athens in the Peloponnesian War). Military success may tempt combatants to increase their aims until they approach the unattainable, while military setbacks may encourage combatants to withdraw prematurely from a conflict, incurring losses that might have been recovered with only modest effort and expense. Rationality provides combatants with a longer-view of the objective, tempering or fuelling the passions necessary to maintain the struggle while tailoring military operations towards the chosen political aim. In some circumstances—again, re-emphasizing the principle of interaction—rational leadership requires the re-definition of political aims to reflect changes in the trinity (exhaustion of will, for example, or unexpected success).

Understanding the nature of the war, however, requires more than just understanding one's *own* trinity—it requires some understanding of the enemy trinity and how to affect it in ways that contribute to reaching your objective. As Sun Tzu points out, knowing both oneself and one's enemy is a decided advantage in war.[13] If an adversary is fighting as part of a coalition, understanding the trinity of the coalition partners can also be useful—attacking a coalition partner may be an effective way of ending a conflict.[14] Ultimately, the more one knows about adversaries, their trinities and their objectives, the more elements of a conflict over which one can exercise control.[15]

States can opt for violence with some sense of public willingness to sustain that violence. They can choose to attack enemy forces, or to try to break the enemy's will, or some combination of both.[16] They can utilize military operations sequentially, with each successive operation capitalizing on the impact of the previous, or cumulatively hoping to create, over time, a complete collapse of some aspect of the enemy political or military system.[17] They may, in particularly difficult or complex wars, do both.

How does this methodology apply to non-state actors and aspiring insurgents? These groups face a daunting problem. As mentioned above, they are inherently weak—they hold no meaningful political power and all they can offer are promises of a better, more just future. Non-state actors face a choice—or at least a range of choices—but also very substantive constraints. Because they are weak, they are unlikely to be able to defeat enemy forces in conventional battle without a protracted period of mobilization.[18] This weakness forces them to focus on enemy will—their target must be the rational and passionate elements of the trinity, since chance and probability are extremely unlikely to work in their favour.

On one end, they can attempt to focus primarily on the enemy—seeking to coerce their adversary through violence. But terror is only a means to an end and one that has many negative consequences. Terror campaigns tend to fail because their actions are viewed as criminal and abhorrent and the population will tend over time to support efforts to stop terrorist campaigns.

At the other end of the spectrum is the kind of committed, revolutionary insurgency advocated by Mao Zedong.[19] Here the focus of the non-state actor is to not only overthrow, but in fact *replace* the ruling regime. This requires a painstaking sequential campaign of building public support, discrediting the enemy regime and constructing over time a shadow regime that governs liberated zones, extracts resources for the greater war effort and eventually creates conventional forces that can defeat the existing regime. These sequential steps—Mao's 'three phases'—are complemented by a cumulative campaign of violence that gradually undermines the legitimacy and authority of the regime.

Most irregular wars fall somewhere between these two poles. Terrorist groups do seek to build political support, but can be grossly unrealistic about the appeal that their combination of ideological fervour and murderous violence may have on even sympathetic populations. Insurgencies often overreach themselves—they may be based on minority ethnic populations that cannot overthrow the existing regime and instead settle for limited autonomy. Revolutions regularly fail to reach their potential—the principle of interaction suggests that a regime that recognizes it is truly threatened by a powerful ideological opponent may well become much more efficient in mobilizing resources and resist the temptation to be a cooperative adversary. However, some revolutionary movements still achieve substantial territorial control—the FARC in Colombia, for example, or the emerging Pakistani Taliban in the Federally Administered Tribal Areas of Pakistan.

A non-state actor's success, therefore, requires flexibility, effectiveness and above all calculation. To enhance its chances of success, the non-state actor must attack the enemy's trinity while building and enhancing its own. When the means used include terrorism—the deliberate targeting of civilians—as almost all irregular wars do, terror must be used as rationally as possible, to maximize effect while minimizing alienation of potential support. As in conventional war, politics must govern the use of force—but because the means used are so much less discreet and so much more alienating, they must be wielded with even greater creativity and skill in order to achieve a group's political aims.

4. The Irish Republican Army: Ends and Means

The IRA and its various splinter groups[20] are heirs to an almost 150 year old republican tradition. The Irish Republican Brotherhood (IRB) was founded in the United States in 1858, with the objective of eliminating British rule in Ireland through violent rebellion and the support of the Irish-American diaspora.[21] The IRB claimed authority as the legitimate government of Ireland until a Republic could be established. The IRB and its successors are violent, revolutionary organizations. They recognize neither the legitimacy of the existing political process nor the possibility of movement towards their goals *without* violence—the legacy of hundreds of years of rebellion, political reform and continued British authority.

The IRB's claim as the only legitimate representative of an idealistic and (to-date) thoroughly notional Irish republic is simultaneously one of the great strengths and tragic flaws of the IRA. On the one hand, legitimacy has been scrupulously passed on, at least in a mythic sense. The rebels of 1916 were led by the IRB and proclaimed an Irish republic based on that legitimacy. The Irish Volunteers and Sinn Fein both swore allegiance to the

Republican ideal in 1919. The defeated IRA remnants of the Irish Civil War swore allegiance to the Second Dail—the last Irish parliament elected *before* the Anglo-Irish Treaty—and rejected any former member who entered legitimate politics in the new Irish Free State. The remaining seven members of the Second Dail transferred the legitimacy of that body back to the IRA in 1938, just before the IRA re-emerged in a series of attacks on England. Sinn Fein, the political party long associated with the illegal fringe of Irish republicanism, swore allegiance to the IRA in 1948. In 1969 and again in 1986, responding to IRA/Sinn Fein decisions allowing members to take seats in the Irish Dail, Tom Maguire—the last unrepentant Republican survivor of the Second Dail—re-transferred that inheritance of legitimacy, first to the Provisional IRA and later to the Continuity IRA.[22] The Real IRA insists that it is the heir to the Republican tradition, because it is composed of the irreconcilable elements of the IRA after the 1997 ceasefire.

This 150 year tradition is crucial to understanding the multi-generational aspect of the IRA. The republican myth is romantic—often futile but always determined resistance by each generation.[23] It is rooted in the long history of Irish rebellion against British rule and legitimized by the passing of the sword from generation to generation.[24] Ireland is a land of long memories and deep hatreds and many who serve in the IRA see it as unfinished business left to them by their ancestors. Republicanism appeals to those who root for the underdog—at least until the ugly truth of Republican violence re-emerges in a new armed campaign—and serves as a reminder not only of the myths and realities of British rule, but also of partition and the discrimination that once dominated Northern Irish society.

The crucial element of the republican tradition is that success requires violence. A secondary motif is the irrelevance, inadequacy and betrayal associated with legitimate political movements—the illegitimacy of any existing government on the island is taken as a given. Those who seek to work for peaceful political change are viewed as naïve or, at best, useful idiots. The IRA allows participation in major elections, provided its candidates then refuse to take their seats on the principle of abstention—attending a Parliament implicitly recognizes the legitimacy of that body to govern.[25] Members of the movement may participate in municipal elections and local political bodies, but must accept the necessity of violence.

This rejection of politics plays an enormous role in IRA failures. The IRA has the relative luxury of operating in two democracies—the Irish republic and the United Kingdom. In each state, opportunities for peaceful democratic change and reform exist. Voters, for obvious reasons, are unlikely to continue voting for candidates who neither represent them nor help them in their everyday lives. As a result, the IRA risks alienating all but the truest of believers, making the creation of a broad base of support very difficult.

In addition, the republican ideology's suspicion of politics and compromise and the commitment to the primacy of violence reverses the Clausewitzian relationship between strategy and policy. Clausewitz argues that policy must always be predominant—allowing for the prospect of redefining objectives and aims in the light of events and making rational calculations about the costs and benefits of continued struggle. This concept does not co-exist easily with the Irish republican ideology, which seeks relatively unlimited aims (the withdrawal of British presence from the island). Violence is the legitimizing and necessary feature of republican struggle. As a result, the gun frequently has precedence over politics—the IRA has launched risky campaigns with little hope of success, because the leadership is composed of ideologically sound fighters with little or no understanding of the nature of the war (an understanding that can only come either through immensely fortunate intuition or through rational calculation).

The core problem for the IRA is threefold. First, the IRA remains militarily weak relative to its primary opponent (the United Kingdom). Since the IRA's objective is very ambitious and its adversaries (both the United Kingdom and the Loyalist population in the North) are tenacious, this means it must be willing to fight a protracted, high damage campaign to break the will of its adversaries over time. Second, the primacy of violence in Republican ideology makes rational calculation about opening new campaigns or adjusting to success and failure in existing campaigns extraordinarily difficult. One key example is the IRA's inability to recognize the political strength of the Loyalist population—which was written off as passive tools of British imperialism until the 1990s. Third, the primacy of violence tends to alienate the IRA's potential bases of support—the Catholic population in the north, the Irish population writ large and the Irish diaspora in the US—unless it can be carefully modulated and directed. For the IRA to succeed at a protracted war with Britain, it *must* have at least the tacit if not the active support of significant elements in Ireland and the USA.

This tension between the pragmatic requirement for modulated violence and an appeal to broader political support and the ideological drive to reject traditional politics, has dominated every IRA debate and profoundly affected IRA campaigns, strategy and tactics.[26] Every new IRA campaign has been affected by concerns about the response of the Irish state, by a perceived change in political conditions and by optimism about the availability of broad-based Irish political sympathy. IRA splits have been dominated by concerns about the approach to the new Irish state, particularly the willingness to take seats in the Irish Dail.[27] The role of Sinn Fein—prominent in recent years—has fluctuated throughout the IRA's history, reflecting the tension between the potential utility of pragmatic politics and the ideological rejection of compromise.[28]

5. The IRA's Failures: the Campaigns of 1939 and 1956

These three problems led directly to the failures of the two IRA mid-century campaigns. In each, the IRA hoped to utilize discreet, targeted violence to coerce the British into leaving the island. They attempted to control the choice of battle space and operate in ways that would not alienate the Irish populations. They did not recognize, however, the determination and cohesiveness of their adversary, the potential interference of the Irish government, or the inordinate weakness of the IRA itself. As a result, in each campaign the IRA faced a massive disconnect between means and ends. It could neither create the conditions to rapidly force the British to change their policies nor create sufficient public and international support to sustain a protracted campaign. In both cases, the direct opposition of the Irish government played an important role in crippling the campaign.

A. The 1939 Campaign

The 1939 campaign was made possible by widespread disillusionment in the IRA and by the mass exodus of republicans to create the Fianna Fail party in the late 1920s.[29] The Army convention of 1938 elected extremists determined to start a campaign in England. This would resurrect the IRA, re-establish its violent credentials and utilize tactics that had not been systematically pursued in the Anglo-Irish War despite the demands of some of the IRA leadership—i.e., deliberate attacks on civilian and economic targets in England to pressure the British public and government.[30]

The campaign was a failure for a number of reasons, but the two most pertinent were almost certainly terrible timing (the bombing campaign began with a 'declaration of war' on 1 January 1939 and continued into the Second World War) and an abject failure to create and manage public support. The timing problem profoundly affected the impact of IRA tactics. Bombing in cities, killing small numbers of civilians, was an atrocity that drew condemnation from all sides rather than British concessions. Once the Second World War started, of course, the damage the IRA could do paled in comparison to German aerial attacks on Britain, eliminating any coercive leverage the IRA might have hoped to gain. The IRA made contact with the Nazis, including some laughable efforts to help German agents penetrate into the UK.[31] This put the Irish government in a very difficult political situation—the Free State was neutral in the conflict and could not be seen acting directly or indirectly in support of Britain's mortal enemy.[32] This problem was exacerbated when the US entered the war and based troops in Northern Ireland.

In addition, the IRA made little effort to secure bases of public support in Northern Ireland or the Free State. The IRA was not represented by any political voice or organization—Sinn Fein was an independent, almost moribund party at this time.[33] The IRA had been made an illegal organization in the 1930s and refused to recognize the Free State.[34] IRA operations in the Free State represented a clear threat to government legitimacy, particularly the arms raid that briefly stripped the Free State's army of almost all of its small arms ammunition.[35] Under these circumstances, it is hardly surprising that the Irish government took very strong repressive measures against the IRA and that the Irish public offered almost no opposition to those measures. In the Loyalist dominated province of Northern Ireland, the IRA attracted slightly more public sympathy, but this was not nearly enough to prevent it from being crippled and subdued by emergency measures. By 1945, the IRA had ceased to exist and had to be re-built from the ground up in the aftermath of the Second World War.[36]

The IRA *did* recover, however—a testament to the residual strength of the Republican ideology in Irish society. The IRA attempted to learn lessons from the previous campaign, as evidenced by two important steps taken in 1948. Recognizing that any new campaign would require a sanctuary, the IRA leadership announced General Order #8, which prohibited the carrying out of military operations in the Irish Free State.[37] While not a recognition of the Irish government's legitimacy, General Order #8 was intended to demonstrate to an Irish government that the IRA was not a threat to public order in the South—only in the 'occupied territory' of Northern Ireland. In addition, the IRA re-absorbed Sinn Fein and made it the official political voice of the Republican movement.[38] This provided a means of political activism—running political prisoners for public office and participating in local elections—and served as a means of communicating with a wider audience than the republican faithful. The revival of Sinn Fein suggested a more pragmatic attitude towards political coalition building.

B. The Border Campaign, 1956-1962

The new IRA campaign, which actually began with a series of arms raids in the early 1950s, was aimed expressly at Northern Ireland.[39] This was a deliberate strategy, reflecting a reconsideration of recent military and political events. First, a new government had declared an Irish republic in the Free State, suggesting a groundswell of public support for republican principles. The possibility of re-uniting North and South, therefore, re-emerged as an issue on both sides of the border.[40] In addition, the IRA rejected a bombing campaign in Britain—partly from preference for a more legitimate military campaign using flying columns and guerrilla raids and partly because of the negative political consequences that the 1939-40 bombings had generated in Ireland and internationally. The IRA 'Handbook for Volunteers' stressed Ireland's guerrilla tradition.[41] Finally, the campaign was expressly designed to be run from sanctuaries in the south, possibly supported by bases established in the more remote border areas in Northern

Ireland.[42] Operations in Belfast were deliberately avoided in an effort to minimize the negative impact on what was considered a hostage population.[43]

Despite careful planning, the Border Campaign proved an utter failure. Again, the primary reasons included timing and the inability to generate public support. The IRA opened the campaign shortly after Sinn Fein candidates received over 150,000 votes on a protest ticket.[44] It was assumed, incorrectly, that this indicated massive support for republican principles in the North, which combined with the republican rhetoric of a Coalition government in Ireland to create favourable conditions for a campaign.[45] In fact, support in the North was only marginal at best. The coalition government in Ireland was defeated in March 1957 elections and the new Fianna Fail government immediately cracked down on the IRA.[46] The campaign fizzled out after about a year, but was not formally ended until 1962.[47] From any reasonable perspective, the IRA had finally burned itself out after a half-century of episodic violence. Shortly after the end of the campaign, the Prime Ministers of Northern Ireland and the Irish Republic opened a new era of political dialogue.

6. IRA Successes: the Anglo-Irish War and the Troubles

Defining either of these campaigns as a success will, of course, raise a series of questions. Republicans of the Irish variety would refer to the Anglo-Irish War (1916-1921) as a failure because it failed to achieve a united Republic.[48] Critics of the notion that terrorism can 'win' might point to the lost opportunities for democratic change in early 20th century Ireland and the host of political problems that years of irregular and civil war caused both North and South of the border. Few would argue that the Troubles (1969-1997) were a success—the IRA failed in its efforts to drive the British from Ireland and were responsible for the deaths of roughly 1500 people and for untold ruin and devastation.

Assessing success or failure purely in terms of maximal or idealistic goals, however, is neither realistic nor practical. Admittedly, the IRA is a revolutionary organization and aspires to more than it accomplished. But absolute victory is rarely attained even in conventional war. As Clausewitz points out, war is a political negotiation waged by force as well as other instruments.[49] When considering partial success for non-state actors, two elements that certainly should be included are whether the conflict ended with the prospects of a lasting and better condition for the non-state combatant and its supporters (a paraphrase of B. H. Liddell Hart),[50] and also the relative condition of the non-state organization *after* the conflict.

A. The Anglo-Irish War (1916-1921)

I have chosen to date this war from the Easter Rising—the first armed insurrection against British rule in Ireland during the 20th century—

and to exclude the bloody Irish Civil War that followed (1922-1923). The Irish republican movement played a crucial role in this conflict, co-opting the leadership of the Irish Volunteers and staging the Easter Rising,[51] assisting the evolution of Sinn Fein as an umbrella political movement incorporating a much broader range of opinion and support,[52] and gradually escalating violence from political assassination to raids on police outposts and eventually to fairly sophisticated rural insurgency.[53]

The Anglo-Irish War represents, in many respects, a proto-typical insurgency. The Irish raised a nationwide resistance against British rule.[54] The rebels established an effective media campaign, utilizing public relations officers and a wide range of contacts in Irish, British and foreign press offices.[55] An alternative government was established, including elected representatives, which provided a substantial range of services including law enforcement, taxation and surprisingly impartial judicial procedures.[56] The Irish 'government' empowered diplomats and ambassadors and sent emissaries to the United States and many leading European powers, as well as to the Versailles conference.[57] The rebels also utilized economic tools, including boycotts of Northern Irish goods in an effort to affect policies in both Britain and Northern Ireland.[58]

Although the Irish rebel government made full use of a wide set of non-military tools, the military campaign it waged against the English was both efficient and relatively precise.[59] The IRA initially focused on the British intelligence network in Ireland.[60] In 1919, the IRA began a sophisticated attack on the Irish police including social ostracization, attacks on isolated police outposts in the countryside and a program of deliberate assassination aimed at the Dublin Metropolitan Police (DMP) 'G Division,' responsible for political intelligence.[61] The IRA effectively infiltrated the Royal Irish Constabulary (RIC) and DMP, to the point that by 1920 British authorities had written them off as a useful source of intelligence.[62]

In 1920, IRA efforts expanded to the countryside, systematically attacking small isolated police outposts and driving the RIC into the towns and cities.[63] By the autumn of 1920, the IRA had established 'flying columns'—permanent guerrilla bands of platoon and company size that patrolled local areas attempting to ambush police and army units.[64] The IRA adapted organizationally, creating larger units across county borders and de-centralizing training and logistics.[65] As British counterinsurgency efforts became more effective, the IRA changed its own patterns to create larger numbers of attacks, generating statistics 'proving' that the British were unsuccessful at containing or rolling back IRA violence.[66] In May 1921, the IRA launched a massive raid against the Customs House in Dublin—visibly contradicting British claims that Dublin was almost under control.[67]

The IRA engaged in and considered brutal acts of terrorism, although the impact of these was limited not only by technology but also by the personal intervention of some IRA leaders. Cathal Brugha, the 'Minister of Defence' in the Irish shadow government, made some of the most outrageous suggestions. He began planning to kidnap the British Cabinet in late 1918 and plans to kidnap Prime Minister Lloyd George reached very advanced stages later in the conflict.[68] Brugha also called for machine gun attacks against British crowds at movie theatres—an effort that was singularly vetoed by Michael Collins.[69] A bombing campaign in Britain devastated the Liverpool docks—with no loss of life—in 1920 and over 1000 attacks were carried out on British soil during 1920-21.[70] British judicial officials were assassinated, generals kidnapped and police coerced. Leaders of the IRA considered the use of biological attacks on British troops and livestock.[71] Finally, the IRA used coercion to keep its own people in line as well as to intimidate and eliminate Loyalists and other potential political opponents.[72]

In the end, the campaign was a singular success. At relatively low cost in lives, the British agreed to negotiate with the Irish representatives and signed a treaty establishing the Irish Free State in December 1921. A quick comparison between the main points of the Home Rule legislation adopted but not enforced in 1914 and the terms of the Treaty of 1921, demonstrates that the IRA achieved much higher levels of sovereignty and independence through its violent efforts than Home Rule would have provided. Home Rule legislation would have provided an Irish parliament purely to consider domestic affairs, leaving foreign affairs in the hands of Westminster. The Free State had Dominion status (like Canada), membership in the League of Nations and the right to govern its own economic and foreign affairs. A crucial stumbling block was the oath of loyalty Irish representatives would have to make to the monarchy—an issue that enraged doctrinaire republicans and contributed to the outbreak of Civil War.

By the standards mentioned earlier—better outcome after the conflict and enhanced position for the organization—the IRA was dramatically successful in the Anglo-Irish War. The IRA achieved far more than even the most ambitious nationalist politicians would have hoped for in 1914. Organizationally, IRA leaders became some of the dominant figures in Irish political life after the establishment of the Free State. The two major political parties in Ireland today—Fine Gael and Fianna Fail—both trace their roots to the opposing sides of the Irish Civil War. IRA leaders who became prominent politicians include Eamon De Valera and Sean Lemass, both of whom eventually became Prime Ministers of Ireland and Sean MacBride, one of the founders of Amnesty International and a recipient of the Nobel Peace Prize.

B. The Troubles, 1969-1997

In 1969, Loyalist rioting and violence in response to a peaceful civil rights campaign in Northern Ireland caused the most massive population shift in Europe since the end of the Second World War. The IRA was unable to defend Catholic areas against Protestant rioters. The Dublin leadership had been deliberately shifting the republican movement towards Marxism and participation in national-level politics in Ireland and the movement lacked both weapons and financial resources.[73] The British Army intervened in August 1969 to provide security for the Catholic population. A special Irish Republican Army Convention held in late 1969 led to a split in the Republican movement—supporters of the Dublin leadership and its decision to end abstention from the Irish Dail constituted the majority of delegates and referred to themselves as the Official IRA. A splinter faction, led primarily by former activists from Northern Ireland, established itself as the Provisional IRA. Within a year, the Provisionals had become the dominant element in the Republican movement and over time asserted themselves as 'the' IRA. The Officials called a ceasefire in 1972 and were absorbed into the political sphere (although they maintained an armed wing which periodically engaged in criminality and violence).[74]

The Provisionals (Provos) began their efforts with a simple strategy. Their first efforts were to position themselves as defenders of the community.[75] Next, they would instigate violence and retaliate against both the police and the British Army and then finally progress to a full armed campaign—bolstered by new recruits from the first two phases.[76] Although the ostensible political objectives of this campaign would most likely be the collapse of Northern Ireland's regional parliament in Stormont and then the withdrawal of British forces,[77] it does not appear that the IRA leadership engaged in extensive thinking about how best to use force to achieve those ends. Instead, a province-wide armed campaign appears to have been an end in itself, with eventual success more a matter of belief than planning.

IRA initial efforts were largely successful. The collapse of the sectarian Northern Irish security forces and Britain's inadequate efforts to reform them placed the security burden on the British Army—a blunt instrument, still obeying the orders of a Northern Irish government that wished to enforce the previous fifty years of discrimination.[78] The British Army's efforts contributed enormously to the legitimization and growth of the IRA, from the 1971 Falls Road curfew, to the decision to reintroduce internment without trial in August 1971, to the tragic and brutal violence of Bloody Sunday in Derry.[79] By 1972, a re-armed IRA with significant public support had established 'no-go' zones in the major cities of Belfast and Derry, declared all-out war on the British Army and began a series of bombing campaigns that devastated Northern Ireland and, on occasion, England.[80]

By July 1972, the IRA had achieved at least one of its potential political objectives. The British government dissolved the Stormont Parliament and met with IRA representatives to discuss the future of the province. The make-up of the IRA negotiating team demonstrated the breadth of the movement—older veterans of previous campaigns mixed with younger, more politically attuned leaders risen from the ranks in the heady days of 1970-71.[81] The talks were a complete and utter failure—the IRA leadership, having no intention of pursuing compromise, simply issued a list of demands (including insistence on British withdrawal) which were utterly unacceptable to the British government.

From 1972-75, the British began to gradually roll back the IRA, imprisoning hundreds of volunteers.[82] After a ceasefire in 1975-76, the younger generation of leadership—mostly Northerners—began discussing a change in strategy.[83] Their new 'Long War' strategy accepted the fact that the rapid victory hoped for by the original Provisional leadership was not possible. The new strategy, developed by Gerry Adams, Martin McGuinness and Ivor Bell (among others) called for a protracted campaign of attrition, mobilizing Irish and international support and creating costs that would be unacceptable to the British public and government over the long run.[84] These leaders helped draft the new IRA 'Green Book'—the manual used for indoctrinating new IRA volunteers—and also became key figures in the IRA leadership and in the resurrected and newly empowered Sinn Fein.[85]

The Long War strategy explicitly recognized the importance of generating a broader network of public support and the limits the republican base faced in sustaining a protracted struggle. Sinn Fein became increasingly involved in local politics, in the campaign against drugs in both the north and the south and in a range of political issues including human rights, prisoner's rights and civilian deaths at the hands of security forces.[86] Sinn Fein declared solidarity with other 'national liberation' movements, including Palestinian groups, the Sandinistas and the Basque ETA.[87] The biggest boost to Sinn Fein's fortunes, in terms of both publicity and in electoral success, came as a result of the Hunger Strikes of 1981, when hunger striker Bobby Sands was elected to Parliament on an abstentionist ticket.[88]

Electoral successes improved Sinn Fein's visibility, while the apparent success of the new strategy pushed the new leaders to the forefront of the republican movement.[89] Adams and McGuinness reportedly became leading members of the IRA Army Council and began to push a policy of 'the bullet and the ballot box'—a simultaneous political campaign and armed struggle.[90] This campaign was costly—elections stressed the IRA budget—and led quickly to another debate over abstention in the Irish republic. In nearly simultaneous IRA and Sinn Fein conventions, the new leadership eliminated abstention in the south—causing a split in Sinn Fein and the resignation of some of the more doctrinaire IRA members.[91]

The change in abstention policy passed, at least in part, because the IRA was promising a renewed military campaign based on Libyan arms.[92] Although this renewed offensive did take place, it was far from decisive. The largest shipment of Libyan arms was intercepted while still en route and many arms caches were discovered by the alarmed Irish police.[93] The IRA was penetrated by informers at increasingly high levels, which led to arrests, targeted killings (most famously in Loughgall and Gibraltar) and executions of suspects by the IRA's own counterintelligence unit—the 'Nutting Squad.'[94] Reports since the 1997 ceasefire appear to broadly substantiate the case that the head of the Nutting Squad—a man codenamed 'Stakeknife'—was actually working for the British Force Reconnaissance Unit.[95]

By the early 1990s, it was clear that while the IRA could not be eliminated, neither could it inflict unacceptable losses that would force the British to withdraw. In both Belfast and Derry, successful attacks became increasingly rare, pushing the burden of the war to the more capable border units.[96] IRA disasters, including the brutal bombing of a Remembrance Day parade in Enniskillen in 1987, reduced the attractiveness of the IRA and Sinn Fein, an effect which was reflected in electoral results.[97]

A new series of bombings in England created unheard of economic effects, devastating not only economic hubs but also the insurance industry.[98] This may have spurred existing indirect contacts between the British government and the IRA. The IRA and Sinn Fein were also working with the Irish government and the Social Democratic and Labour Party to try to find common diplomatic ground, even if it meant giving up the armed struggle.[99] After torturous multi-party negotiations, an intervention by the US government and a renewed spate of Loyalist murders that put the IRA's reputation as a protector of the Catholic community at risk, a compromise was reached and the IRA declared a ceasefire in 1994. The peace process weathered a collapse in the ceasefire in 1996, an IRA emergency Army convention that almost deposed the Adams-McGuinness leadership and a split in the IRA that created an ineffectual but still dangerous and ideologically pure 'Real IRA.'[100] The peace process remains in place today, including the effective disarming of the IRA and the acceptance of a new Police Service in Northern Ireland that has replaced the Royal Ulster Constabulary *and* the IRA (in nationalist communities) in a local policing role.[101]

Did the IRA 'succeed' in the Troubles? Again, based on the criteria mentioned above, the answer must be a qualified 'yes.' The IRA overthrew the Stormont regime and one result of the extended campaign was the much more rigorous enforcement of British codes on discrimination in the Northern Irish province. The security forces have been reformed and the British government has committed itself to accepting the decision of the Northern Irish population regarding its continued attachment to the United Kingdom.

That is only part of what the leadership of the Provisional IRA hoped to achieve in 1972, but it is a significant partial success. It could, however, have been achieved at any point after 1975 and was largely inevitable after Britain's imposition of direct rule and, by that action, of British law.

Equally important, from the perspective of the revolutionaries, the IRA and their Sinn Fein cohorts (and it would be a mistake to assume too much distance between the two organizations) are now the largest political party representing the interests of the Catholic population in Northern Ireland. They are also the only party that participates in both Irish and UK politics. This is an enormous change in influence for the republican movement. They have achieved both personal and institutional power and wealth by overturning decades of tradition and compromising on matters of highest principle. The question of whether the accomplishments were worth the cost—not only in lives, but for both communities over a nearly 30 year conflict—is not easy to assess. From the IRA's perspective, however, the answer is probably a qualified 'yes.'

7. Conclusions

What does the IRA experience suggest about the ability of non-state actors to win in irregular struggles? First, it demonstrates again that quick victory is unlikely. The mismatch in resources between a non-state actor and a government is such that only a truly cooperative or pathetic government will collapse in a hurry. Protracted campaigns, therefore, are the most likely opportunity for substantial success.

This leads to the second point: the IRA has been successful only when it capitalizes on political causes beyond its own republican ideology. The Anglo-Irish War was made possible by creating an umbrella movement that encompassed those irate over Britain's treatment of the 1916 rebels, those opposing Britain's efforts to apply conscription to Ireland and an increasingly large portion of the constitutional nationalists frustrated at Britain's lack of response to legitimate calls for Irish self-governance. The Provos emerged as the result of legitimate political challenges to the deliberately discriminatory Stormont regime and the brutal repression that Stormont and the Loyalist population unleashed in response. Sinn Fein and the IRA were revived by the martyrdom of ten hunger strikers—a powerful and emotional blow to British legitimacy in the eyes of the Irish and American populations as well as to the nationalists in Northern Ireland. It appears, therefore, that building and maintaining public support both at home and abroad allows revolutionary movements to escalate and protract conflict, raising the possibility of success through attrition or negotiation. In addition, a sense of grievance and injustice plays a major role in the continuity of the movement, even after catastrophic defeat.

A third point is that non-state actors often require the ability to make political compromises, which permits partial victory in the short-term. When the IRA has become ideologically hidebound, as in 1972, it has been unable to accept 'half a loaf.' The civil war of 1922-23 was the result of an inability by some republicans to accept 85% of a loaf—the Treaty might easily be compared to the options the Palestinian delegation faced at Wye River and Camp David. Throughout most of the 20th century, Sinn Fein was completely dominated by the wishes of the IRA leadership. Since that leadership believed in the primacy of violence over negotiation, it made Sinn Fein useful primarily as a propaganda wing. The current IRA/Sinn Fein leadership—which only supported political compromise after a decade-long period of internal coalition building—achieved substantial gains, both from the perspective of the party as an institution and in terms of the initial demands of the IRA and nationalist community in 1969, albeit at enormous and tragic costs in human lives and social disruption.[102]

A final point is the necessity to dominate the interaction between the opposing sides. The IRA has been able to capitalize on and sometimes even provoke, enemy mistakes—the conscription debate in 1918, the Restoration of Order in Ireland Act of 1920 (which put volunteers 'on the run' and provided the manpower for the flying columns), the overreaction of British Army forces in 1970-72, Bloody Sunday and the Hunger Strikes. They have also made great use of timing and serendipity—the 1918 election which created a vast Sinn Fein majority was at least partially fuelled by a change in British electoral laws permitting women to vote for the first time. The IRA has been weakened or failed when it has made catastrophic public failures—Enniskillen, major civilian casualties in urban bombing in Belfast and England and the Real IRA's slaughter at Omagh. The bullet and ballot box strategy ultimately proved the inherent tension in waging a simultaneous legitimate political campaign and an illegal armed struggle in a democratic environment. The IRA cannot control the actions of the security forces, but it can provoke them. On the other hand, the political costs of indiscriminate or errant acts of violence have proven to be extraordinarily high, particularly when the adversary manages to control his own responses.

To return to the basics of strategy, the IRA has been most successful when it has done three things. First, when it has been able to construct broad public support—the 'passion' end of the trinity. Second, when it has been able to modulate its violence and take advantage of enemy errors—the 'chance and probability' end of the trinity. Finally, when it can assert control over its own agenda and accept a 'good but not perfect' solution—the 'rational' end of the trinity. Faced with a gross disproportion between ends and means, the IRA has sometimes been successful by protracting the conflict, carefully managing the violence and accepting reasonable achievements that lead to a better condition.

How might this apply to the current war on terror? Al Qaeda may be building the mystique necessary to maintain a multi-generational movement. Clearly, preventing it from becoming an umbrella political movement would be the most effective way of reducing its influence. Taking advantage of its atrocities—which are, sadly, all too regular—is an opportunity for sophisticated information operations that we do not appear to have capitalized on. Limiting our own errors is clearly a priority—and something the US has been more aware of after Abu Ghraib. Nevertheless, as the IRA demonstrates, errors can resonate for years if not decades. Finally, Al Qaeda requires international sympathy and support and some level of sanctuary from friendly or uncommitted regimes. This can be combated more easily through diplomacy and law enforcement cooperation than through overt military strikes—although the occasional act of hot pursuit may be necessary. Al Qaeda wants to create a broadly based insurgency. The most likely way to do that is to provoke US military over-reaction—a temptation we should go to great pains to avoid.

Notes

[1] See, for example, C. Carr, *The Lessons of Terror*, Random House, New York, 2002.
[2] Sun Tzu, *The Art of War*, with an introduction by S.B. Griffith (trans), with a foreword by B.H. Liddell-Hart, Oxford University Press, Oxford, 1971; and C. von Clausewitz, *On War*, Indexed Edition, M. Howard & P. Paret (ed), Introductory Essays by P. Paret, M. Howard, & B. Brodie, with a Commentary by B. Brodie, Princeton University Press, Princeton, 1976.
[3] *Quadrennial Defense Review Report*, Department of Defense, Washington, 6 February 2006, Available at the Commonwealth Institute website, Accessed 3 September 2008, <http://www.comw.org/qdr/qdr2006.pdf>.
[4] *On War* contains a chapter entitled 'The People in Arms.' Clausewitz, op. cit., Book VI, Chapter 26, pp. 479-483.
[5] T.G. Mahnken, 'Why the Weak Win: Strong Powers, Weak Powers and the Logic of Strategy,' *Strategic Logic and Political Rationality: Essays in Honor of Michael I. Handel,* B.A. Lee & K.F. Walling (eds), Frank Cass, London, 2003, pp. 60-73; T.V. Paul, *Asymmetric Conflicts: War Initiation by Weaker Powers*, Cambridge University Press, Cambridge, 1994; and M. Handel, *Weak States in the International System,* Frank Cass, London, 1990.
[6] For a comparative discussion of rationality in both Sun Tzu and Clausewitz, see M. Handel, *Masters of War: Classical Strategic Thought, 3rd Expanded and Revised Edition*, Frank Cass, London, 2001, pp. 77-80. Clausewitz also emphasizes the unpredictability of war, which is at least in part the result of interaction with an adversary who will neither necessarily cooperate nor

necessarily behave in ways that will appear rational to other combatants. Clausewitz, op. cit., pp. 139, 161.
[7] 'In war, the will is directed at an animate object that *reacts*.' Clausewitz, op. cit., p. 149.
[8] Clausewitz, op. cit., pp. 90-99. Clausewitz devotes an entire book (Book 6) to the study of the defense that deserves greater study. See pp. 357-519. Sun Tzu counsels against allowing a war to become protracted. Sun Tzu, op. cit., Book II, 3-7, p. 73.
[9] Clausewitz, op. cit., p. 358.
[10] Sun Tzu, op. cit., Book I, pp. 63-71, 85-95. Clausewitz, however, also notes that 'Since war is not an act of senseless passion but is controlled by its political object, the value of this object must determine the sacrifices to be made for it in *magnitude* and also in *duration*. Once the expenditure of effort exceeds the value of the political object, the object must be renounced and peace must follow.' Clausewitz, op. cit., p. 92.
[11] 'No one starts a war—or rather, no one in his sense ought to do so—without first being clear in his mind what he intends to achieve by that war and how he intends to conduct it.' Clausewitz, op. cit., p. 579.
[12] Clausewitz, op. cit., p. 89.
[13] 'Know the enemy and know yourself: in a hundred battles you will never be in peril.' Sun Tzu, op. cit., p. 84.
[14] Clausewitz, op. cit., p. 596; Sun Tzu, op. cit., p. 78.
[15] J.C. Wylie has proposed that 'The primary aid of the strategist in the conduct of war is some selected degree of control of the enemy for the strategist's own purpose; this is achieved by control of the pattern of war; and this control of the pattern of war is had by manipulation of the center of gravity of war to the advantage of the strategist and the disadvantage of the opponent.' J.C. Wylie, *Military Strategy: A General Theory of Power Control*, Rutgers University Press, New Brunswick, 1967, p. 91.
[16] Clausewitz, op. cit., p. 77.
[17] Wylie, op. cit., pp. 23-29.
[18] M. Zedong (Tse-Tung), *On Guerrilla Warfare*, with an introduction by Brigadier General S.B. Griffith (trans), Praeger, New York, 1961.
[19] M. Zedong, 'On Protracted War' and 'The Present Situation and Our Tasks,' *Selected Writings of Mao Tse-tung*, Foreign Languages Press, Peking, 1967, pp. 206-263, 345-352.
[20] These include the Official IRA, the Irish National Liberation Army, the Continuity IRA and the Real IRA.
[21] O. McGee, *The IRB: The Irish Republican Brotherhood from the Land League to Sinn Fein*, Four Courts Press, Dublin, 2005.

[22] R. English, *Armed Struggle: The History of the IRA*, Oxford University Press, Oxford, 2003, pp. 113-114.
[23] This concept of 'apostolic succession' is discussed in M.L.R. Smith, *Fighting for Ireland? The Military Strategy of the Republican Movement*, Routledge, London, 1995, pp. 11-12.
[24] The Proclamation of 1916 specifically mentions that 'In every generation the Irish people have asserted their right to national freedom and sovereignty; six times during the last three hundred years they have asserted it to arms.' See also J.B. Bell, *The IRA 1968-2000: Analysis of a Secret Army*, Frank Cass, London, 2000.
[25] This is not the case currently, except in the more dogmatic splinter factions like the Real IRA and Continuity IRA.
[26] B. Feeney, *Sinn Fein: A Hundred Turbulent Years*, University of Wisconsin Press, Madison, WI, 2002.
[27] The two most famous recent divisions of the IRA occurred in 1969-70—which led to the formation of the Provisional IRA—and again in 1986. See J. B. Bell, *The Secret Army: The IRA, Revised 3rd Edition*, Transaction Publishers, New Brunswick, NJ, 1997, pp. 355-372; H. Patterson, *The Politics of Illusion: A Political History of the IRA*, Serif, London, 1997, pp. 96-151; T.P. Coogan, *The IRA, Fully Revised and Updated*, Palgrave, New York, 2002, pp. 330-338.
[28] The role of Sinn Fein in the Republican movement, particularly since 1970, is examined in P. Taylor, *Behind the Mask: The IRA and Sinn Fein*, TV Books, New York, 1997; and E. Moloney, *A Secret History of the IRA*, W. W. Norton, New York, 2002.
[29] The period from 1923, when the IRA ended its war against the new Irish Free State and 1946, when the IRA was apparently crushed, is discussed in C. Foley, *Legion of the Rearguard: The IRA and the Modern Irish State*, Pluto Press, London, 1992; and U. MacEoin, *The IRA in the Twilight Years 1923-1948*, Argenta, Dublin, 1997.
[30] Foley, op. cit., pp. 173-187 and Bell, *The Secret Army*, op. cit., pp. 145-156.
[31] S. O'Callaghan, *The Jackboot in Ireland*, Roy Publishers, New York, 1958; Coogan, *The IRA*, pp. 203-217.
[32] See R. Fisk, *In Time of War: Ireland, Ulster and the Price of Neutrality*, Paladin Books, London, 1985.
[33] Feeney, op. cit., pp. 161-183.
[34] Foley, op.cit., pp. 151-158; Coogan, *The IRA*, p. 132.
[35] Coogan, *The IRA*, op. cit., pp. 135-136.
[36] Bell, *The Secret Army*, op. cit., pp. 239-254; Coogan, *The IRA*, op. cit., pp. 253-262.

[37] English, op. cit., p. 71.
[38] Feeney, op. cit., pp. 185-194.
[39] Bell, *The Secret Army*, op. cit., pp. 250-271; Coogan, *The IRA*, op. cit., pp. 256-276.
[40] Feeney, op. cit., pp. 184-205.
[41] *Handbook for Volunteers of the Irish Republican Army: Notes on Guerrilla Warfare*, Paladin Press, Boulder, CO, 1985.
[42] Bell, *The Secret Army*, op. cit., pp. 280-294; Coogan, *The IRA*, op. cit., p. 302; English, op. cit., p. 73.
[43] Coogan, *The IRA*, op. cit., p. 302.
[44] Feeney, op. cit., p. 202; Bell, *The Secret Army*, op. cit., p. 277.
[45] Feeney, op. cit., pp. 203-210.
[46] Bell, *The Secret Army*, op. cit., pp. 305-7; Feeney, op. cit., pp. 205-6; English, op. cit., p. 74.
[47] Brief histories of the campaign can be found in Coogan, *The IRA*, op. cit., pp. 297-329; Bell, *The Secret Army*, op. cit., pp. 289-336.
[48] See in particular D. Macardle, *The Irish Republic*, Farrar, Straus & Giroux, New York, 1999.
[49] Clausewitz, op. cit., p. 87. Later, however, he emphasizes that '…war itself does not suspend political intercourse or change it into something entirely different. In essentials that intercourse continues irrespective of the means it employs…' Clausewitz, op. cit., p. 605.
[50] 'The object in war is to attain a better peace—even if only from your own point of view.' B.H. Liddell Hart, *Strategy, Second Revised Edition*, Meridian, New York, 1991, p. 353.
[51] See J. Stephens, *The Insurrection in Dublin*, Colin Smythe, Ltd., Buckinghamshire, 1992; M. Caulfeld, *The Easter Rebellion: Dublin 1916*, Roberts Rinehart Publishers, Boulder, CO, 1995; C. Townshend, *Easter 1916: The Irish Rebellion*, Allen Lane, London, 2005.
[52] P.S. O'Hegarty, *The Victory of Sinn Fein*, University College Dublin Press, Dublin, 1998; M. Laffan, *The Resurrection of Ireland: The Sinn Fein Part 1916-1923*, Cambridge University Press, Cambridge, 1999.
[53] T.R. Dwyer, *The Squad and the Intelligence Operations of Michael Collins*, Mercier Press, Cork, 2005; M.T. Foy, *Michael Collins' Intelligence War*, Sutton Publishing, Gloucestershire, 2006; J. Augusteijn, *From Public Defiance to Guerrilla Warfare*, Irish Academic Press, Dublin, 1996.
[54] M. Hopkinson, *The Irish War of Independence*, McGill-Queen's University Press, Montreal, 2002.
[55] K. Inoue, 'Propaganda II: Propaganda of Dail Eireann, 1919-21,' *The Irish Revolution 1913-1923*, J. Augusteijn (ed), Palgrave, New York, 2002, pp. 87-

102; 'Publicity Memo #1.' GHQ Dublin, 4 May 1921, *Mulcahy Papers*, University College Dublin, A18.
[56] A. Mitchell, 'Alternative Government', *The Irish Revolution 1913-1923*, op. cit., pp. 70-86; F. Costello, *The Irish Revolution and its Aftermath, 1916-1923*, J. Augusteijn (ed), Irish Academic Press, Dublin, 2003, pp. 186-204.
[57] F. Costello, op. cit., p. 44; National Archives Ireland, Dublin, Bureau of Military History WS790, 'Statement by Robert Brennan.'
[58] Macardle, op. cit., pp. 387, 659.
[59] C .Townshend, *Political Violence in Ireland: Government and Resistance since 1848*, Clarendon Press, Oxford, 1983, pp. 332-340; Smith, op. cit., pp. 30-37; C. Gray, 'The Anglo-Irish War, 1919-1921: Lessons from an Irregular Conflict,' *Comparative Strategy*, 2007, pp. 371-374.
[60] Hopkinson, op. cit., pp. 69-78; Dwyer, op. cit.; Foy, op. cit.
[61] Five G division detectives were killed and one permanently disabled in 1919. See Hopkinson, op. cit., p. 55. For a lengthy discussion of the makeup of the Dublin Metropolitan Police and its G Division, see Bureau of Military History 1280, 'Statement of Eamonn Broy,' pp. 15-25.
[62] C. Townshend, *The British Campaign in Ireland 1919-1921: The Development of Political and Military Policies*, Oxford University Press, Oxford, 1975, pp. 41-42.
[63] Hopkinson, op. cit., pp. 69-73.
[64] T. Barry, *Guerrilla Days in Ireland*, Anvil Books, Dublin, 1990, pp. 19-20; T.P. Coogan, *The Man Who Made Ireland*, Roberts Rinehart Publishers, Niwot, CO, 1992, p. 140.
[65] M.G. Valiulis, *Portrait of a Revolutionary: General Richard Mulcahy and the Founding of the Irish Free State*, The University Press of Kentucky, Lexington, KY, 1992, pp. 70-73.
[66] Townshend, *The British Campaign in Ireland 1919-1921*, op. cit., pp. 185, 214.
[67] Macardle, op. cit., pp. 462-463.
[68] Coogan, *The Man Who Made Ireland*, op. cit., pp. 179, 183.
[69] Ibid., p. 179.
[70] Ibid., p. 153; P. Hart, *The I.R.A. at War 1916-1923*, Oxford University Press, Oxford, 2003, pp. 141-177.
[71] Foy, op. cit., pp. 127-128.
[72] P. Hart, *The I.R.A. & Its Enemies: Violence and Community in Cork, 1916-1923*, Clarendon Press, Oxford, 1998.
[73] Coogan, *The IRA*, op. cit., pp. 330-339; Bell, *The Secret Army*, op. cit., pp. 337-352.
[74] J.B. Bell, *The IRA: Tactics and Targets*, Poolbeg, Dublin, 1990; Bell, *The Secret Army*, op. cit., pp. 356-392; Smith, op. cit., pp. 89-90.

[75] Bell, *The Secret Army*, op. cit., p. 376; Coogan, *The IRA*, op. cit., pp. 365, 437; English, op. cit., pp. 134-5.
[76] S. MacStiofain, *Revolutionary in Ireland*, R&R Clark, Ltd., Edinburgh, 1975. The author was the first Provisional IRA Chief of Staff.
[77] See Smith, op. cit., pp. 90-105.
[78] R. Thornton, 'Getting it Wrong: The Crucial Mistakes Made in the Early Stages of the British Army's Deployment to Northern Ireland (August 1969 to March 1972),' *The Journal of Strategic Studies*, Vol. 30, February 2007, pp. 73-107.
[79] Moloney, op. cit., pp. 74-110.
[80] Romantic descriptions of these developments can be found in MacStiofain, op. cit., pp. 225-269; K. Kelley, *The Longest War: Northern Ireland and the IRA*, Brandon Books, Dingle, 1982, pp. 120-262. Regarding attacks in Britain, both are overt apologists for the IRA. Coogan, *The IRA*, op. cit., pp. 385-391 and Bell, *The Secret Army*, op. cit., pp. 411-416 are more measured, while Smith, op. cit., pp. 123-128 has more details on the breadth of the British campaign.
[81] MacStiofain, op. cit., pp. 271-289. P. Bishop & E. Mallie, *The Provisional IRA*, William Heinemann Ltd., London, 1987, pp. 164-179; and Taylor, op. cit., pp. 153-174, provide insight into the IRA's negotiating team.
[82] English, op. cit., pp. 161-183; Smith, op. cit., pp. 128-142
[83] B. O'Brien, *The Long War: The IRA and Sinn Fein 1985 to Today*, Syracuse University Press, Syracuse, NY, 1995, pp. 103-117; Smith, op. cit., pp. 143-168.
[84] Moloney, op. cit., pp. 149-162; Taylor, op. cit., pp. 245-253, Bishop & Mallie, op. cit., pp. 246-268.
[85] Extracts from the Green Book can be found in O'Brien, op. cit., pp. 351-355. The use of the Green Book as both a teaching tool and an indoctrination process is described in E. Collins, *Killing Rage*, Granta Book, London, 1997, p. 66.
[86] Feeney, op. cit., pp. 251-291.
[87] Kelley, op. cit.; MacStiofain, op. cit.; and Moloney, op. cit., pp. 180-189.
[88] Bishop, op. cit., pp. 187-226. Extensive studies of the IRA's prison struggle include T.P. Coogan, *On The Blanket: The H-Block Story*, Ward River Press, Dublin, 1980; and D. Beresford, *Ten Men Dead*, Grafton Books, London, 1987. See also Moloney, Ibid., pp. 189-190 and 205-215.
[89] Moloney, op. cit., pp. 238-245.
[90] Feeney, op. cit., pp. 292-333.
[91] Moloney, op. cit., pp. 287-297; English, op. cit., pp. 249-253; O'Brien, op. cit., pp. 118-132.

[92] Moloney, op. cit., p. 292; English, op. cit., p. 249; O'Brien, op. cit., pp. 135-141.
[93] O'Brien, op. cit., pp. 142-153; Moloney, op. cit., pp. 3-24.
[94] Moloney, op. cit., pp. 24-33, 304-318, 326-331.
[95] M. Ingram & G. Harkin, *Stakeknife: Britain's Secret Agents in Ireland,* The University of Wisconsin Press, Madison, 2004; S O'Callaghan, *The Informer*, Bantam Press, London, 1998; 'Sinn Fein's Top Secret Spy,' *The Sunday World*, 28 May 2006, Accessed 18 September 2008, available at: <http://cryptome.org/mcguinness-spy.htm> .
[96] See T. Harnden, *'Bandit Country': The IRA and South Armagh, Revised and Updated*, Hodder & Stoughton, London, 1999; Moloney, op. cit., pp. 318-319.
[97] Smith, op. cit., pp. 175-178; Taylor, op. cit., p. 324; Feeney, op. cit., pp. 339-342.; O'Brien. op. cit., pp. 142-143, 151-152.
[98] O'Brien, op. cit., p. 272; Taylor, op. cit., pp. 377-378; Harnden, op. cit., pp. 5-10, 21-27, 313-351.
[99] O'Brien, op. cit., pp. 224-262; Feeney, op. cit., pp. 334-428.
[100] The Real IRA was substantially crippled after the disastrous Omagh bombing in 1998. See J Mooney & M O'Toole, *Black Operations: The Secret War Against the Real IRA,* Maverick House, Ashbourne, 2004.
[101] On decommissioning of IRA arms stockpiles, see 'Report of the Independent International Commission on Decommissioning,' 19 January 2006, Accessed 18 September 2008, available at: <http://www.justice.ie/en/JELR/IICDletter.pdf/Files/IICDletter.pdf>. Other documents of the commission pertaining to policing and normalization of security can be found at <http://www.justice.ie/en/JELR/Pages/Publications_Northern_Ire land>.
[102] Moloney, op. cit., is the most exhaustive account of this process currently available.

Bibliography

Augusteijn, J., *From Public Defiance to Guerrilla Warfare.* Irish Academic Press, Dublin, 1996.

Barry, T., *Guerrilla Days in Ireland.* Anvil Books, Dublin, 1990.

Bell, J.B., *The IRA 1968-2000: Analysis of a Secret Army.* Frank Cass, London, 2000.

—, *The IRA: Tactics and Targets.* Poolbeg, Dublin, 1990.

—, *The Secret Army: The IRA, Revised 3rd Edition*. Transaction Publishers, New Brunswick, NJ, 1997.

Beresford, D., *Ten Men Dead*. Grafton Books, London, 1987.

Bishop, P. & Mallie, E., *The Provisional IRA*. William Heinemann Ltd., London, 1987.

Carr, C., *The Lessons of Terror*. Random House, New York, 2002.

Caulfeld, M., *The Easter Rebellion: Dublin 1916*. Roberts Rinehart Publishers, Boulder, CO, 1995.

Collins, E. *Killing Rage*. Granta Books, London, 1997.

Coogan, T.P., *On The Blanket: The H-Block Story*. Ward River Press, Dublin, 1980.

—, *The IRA, Fully Revised and Updated*. Palgrave, New York, 2002.

—, *The Man Who Made Ireland*. Roberts Rinehart Publishers, Niwot, CO, 1992.

Costello, F., *The Irish Revolution and its Aftermath, 1916-1923*. Irish Academic Press, Dublin. 2003.

Dwyer, T.R., *The Squad and the Intelligence Operations of Michael Collins*. Mercier Press, Cork, 2005.

English, R., *Armed Struggle: The History of the IRA*. Oxford University Press, Oxford, 2003.

Feeney, B., *Sinn Fein: A Hundred Turbulent Years*. University of Wisconsin Press, Madison, WI, 2002.

Foley, C., *Legion of the Rearguard: The IRA and the Modern Irish State*. Pluto Press, London, 1992.

Foy, M.T., *Michael Collins' Intelligence War*. Sutton Publishing, Gloucestershire, 2006.

Gray, C., 'The Anglo-Irish War, 1919-1921: Lessons from an Irregular Conflict.' *Comparative Strategy*. Vol. 26, 2007, pp. 371-374.

Handbook for Volunteers of the Irish Republican Army: Notes on Guerrilla Warfare. Paladin Press, Boulder, CO, 1985.

Handel, M., *Masters of War: Classical Strategic Thought, 3rd Expanded and Revised Edition*. Frank Cass, London, 2001.

—, *Weak States in the International System*. Frank Cass, London, 1990.

Harnden, T., *'Bandit Country': The IRA and South Armagh, Revised and Updated*. Hodder & Stoughton, London, 1999.

Hart, P., *The I.R.A. & Its Enemies: Violence and Community in Cork, 1916-1923*. Clarendon Press, Oxford, 1998.

—, *The I.R.A. at War 1916-1923*. Oxford University Press, Oxford, 2003.

Hopkinson, M., *The Irish War of Independence*. McGill-Queen's University Press, Montreal, 2002.

Ingram, M. & Harkin, G., *Stakeknife: Britain's Secret Agents in Ireland*. The University of Wisconsin Press, Madison, 2004.

Inoue, K., 'Propaganda II: Propaganda of Dail Eireann, 1919-21.' *The Irish Revolution 1913-1923*. Augusteijn, J. (ed), Palgrave, New York, 2002.

Kelley, K., *The Longest War: Northern Ireland and the IRA*. Brandon Books, Dingle, 1982.

Laffan, M., *The Resurrection of Ireland: The Sinn Fein Part 1916-1923*. Cambridge University Press, Cambridge, 1999.

Liddell Hart, B.H., *Strategy, 2nd Revised Edition*. Meridian, New York, 1991.

Macardle, D., *The Irish Republic*. Farrar, Straus & Giroux, New York, 1965.

MacEoin, U., *The IRA in the Twilight Years 1923-1948*. Argenta, Dublin, 1997.

McGee, O., *The IRB: The Irish Republican Brotherhood from the Land League to Sinn Fein.* Four Courts Press, Dublin, 2005.

MacStiofain, S., *Revolutionary in Ireland.* R&R Clark, Ltd., Edinburgh, 1975.

Mahnken, T.G., 'Why the Weak Win: Strong Powers, Weak Powers and the Logic of Strategy.' *Strategic Logic and Political Rationality: Essays in Honor of Michael I. Handel.* Lee, B.A. & Walling, K.F. (eds), Frank Cass, London, 2003.

M. Zedong (Tse-Tung), *On Guerrilla Warfare.* With an Introduction by Brigadier General S.B. Griffith, Praeger, New York, 1961.

—, 'On Protracted War.' *Selected Writings of Mao Tse-tung.* Foreign Languages Press, Peking, 1967.

—, 'The Present Situation and Our Tasks.' *Selected Writings of Mao Tse-tung.* Foreign Languages Press, Peking, 1967.

Mitchell, A., 'Alternative Government.' *The Irish Revolution 1913-1923.* Augusteijn, J. (ed), Palgrave, New York, 2002.

Moloney, E., *A Secret History of the IRA.* W. W. Norton, New York, 2002.

Mooney, J. & O'Toole, M., *Black Operations: The Secret War against the Real IRA.* Maverick House, Ashbourne, 2004.

O'Brien, B., *The Long War: The IRA and Sinn Fein 1985 to Today.* Syracuse University Press, Syracuse, NY, 1995.

O'Callaghan, S., *The Informer.* Bantam Press, London, 1998.

O'Callaghan, S., *The Jackboot in Ireland.* Roy Publishers, New York, 1958.

O'Hegarty, P.S., *The Victory of Sinn Fein.* University College Dublin Press, Dublin, 1998.

Patterson, H., *The Politics of Illusion: A Political History of the IRA.* Serif, London, 1997.

Paul, T.V., *Asymmetric Conflicts: War Initiation by Weaker Powers.* Cambridge University Press, Cambridge, 1994.

'Publicity Memo #1,' GHQ Dublin, 4 May 1921, *Mulcahy Papers.* University College Dublin, A18.

Quadrennial Defense Review Report. Department of Defense, Washington, 6 February 2006, Available at the Commonwealth Institute website. Accessed 3 September 2008, Available at: <http://www.comw.org/qdr/qdr 2006.pdf>.

'Report of the Independent International Commission on Decommissioning.' 19 January 2006, accessed 18 September 2008, available at: <http://www.justice.ie/en/JELR/IICDletter.pdf/Files/IICDletter.pdf >.

'Sinn Fein's Top Secret Spy.' *The Sunday World.* 28 May 2006, accessed 18 September 2008, available at: <http://cryptome.org/mcguinness-spy.htm>.

Smith, M.L.R., *Fighting for Ireland? The Military Strategy of the Republican Movement.* Routledge, London, 1995.

Stephens, J., *The Insurrection in Dublin.* Colin Smythe, Ltd., Buckinghamshire, 1992.

S. Tzu, *The Art of War.* With an Introduction by Brigadier General S.B. Griffith (trans.), with a Foreword by B.H. Liddell Hart. Oxford University Press, Oxford, 1971.

Taylor, P., *Behind the Mask: The IRA and Sinn Fein.* TV Books, New York, 1997.

Thornton, R., 'Getting it Wrong: The Crucial Mistakes Made in the Early Stages of the British Army's Deployment to Northern Ireland (August 1969 to March 1972).' *The Journal of Strategic Studies.* Vol. 30, No. 1 (February 2007), pp. 73-107.

Townshend, C., *Easter 1916: The Irish Rebellion.* Allen Lane, London, 2005.

—, *Political Violence in Ireland: Government and Resistance since 1848.* Clarendon Press, Oxford, 1983.

—, *The British Campaign in Ireland 1919-1921: The Development of Political and Military Policies.* Oxford University Press, Oxford, 1975.

Valiulis, M.G., *Portrait of a Revolutionary: General Richard Mulcahy and the Founding of the Irish Free State.* The University Press of Kentucky, Lexington, KY, 1992.

Wylie, J. C., *Military Strategy: A General Theory of Power Control.* Rutgers University Press, New Brunswick, 1967.

Timothy D. Hoyt is a Professor of Strategy and Policy and has recently been named John Nicholas Brown Chair of Counterterrorism Studies at the United States Naval War College. He served in Belfast, Northern Ireland as a volunteer with the Ulster Quaker Service Committee in 1982 and 1983 and maintains an enormous affection for Ireland north and south.

Tiger, Tiger Burning Bright

Graeme Goldsworthy, Toby Chesson and Erica Pasini

Abstract
The International Campaign to Ban Landmines (ICBL) has been the driving force in the creation of an unprecedented international treaty aimed at outlawing the sale, transport and employment of this most insidious weapon (and at very least publically stigmatizing involvement with it). Since 1997, 156 countries have signed or acceded to the treaty. Nevertheless the effects of landmine use throughout the world's conflict zones have not dropped dramatically. Nations and NGOs involved in this process are now questioning the methods and over-arching logic used to justify the Mine Ban Treaty. This chapter concentrates upon the armed conflict in Sri Lanka, which has been raging for more than two decades. It makes the case that due to the very nature of this civil war—in which suicide bombing and extreme ethnic violence have often gone hand-in-hand with negotiations and have resulted in the Ceasefire Agreement (CFA)—the issue of humanitarian de-mining and international pressure to limit the use of Anti-personnel and Anti-Tank/Vehicle mines will have little effect and while the conflict continues the level of violence affecting non-combatants will remain unacceptably high. More broadly, it will be argued that the transformation in the conduct of low-intensity warfare, especially with respect to ethnic and religious conflicts as well as territorial disputes—which engender the use of mass casualty suicide attacks by guerrillas and insurgents as a strategic weapon—renders moot the ethical arguments applied against these same groups' use of landmines. The treaty, in short, will have little impact in the most typical contemporary conflicts.

Key Words: Landmines, Sri Lanka, ICBL, human security, Ottawa Treaty.

*****

1. The Ottawa Treaty in Practice

It is an unquestioned fact that landmines and explosive remnants of war (ERW) cause horrific injury, often to victims who have little or no grasp of the reasons why the weapons were originally planted and who do not even see the soldier, airman, guerrilla fighter, or child who has wrought such destruction upon their lives and those of their relatives and dependants. Unfortunately, this description captures a raw reality of modern warfare or low intensity conflict (LIC) and it is in this vein that this chapter shall assert that the effective functionality of the anti-personnel mine (APM) and other similar weapon systems will continue to take priority over the moral and even legal arguments to ban these weapons.

In attempting to create a legal ban on so prolific a weapon and one which has come to define the way that many intra-state conflicts are fought in the twentieth century,[1] the Ottawa Treaty, spearheaded by the ICBL, has run into a fundamental difficulty in presenting what amounts to a *fait accompli* to the international community (including armed sub-state groups many of whom rationalise their use of the APM as a battlefield necessity). Moreover, the vague boundaries of the treaty and the blurred definition of what exactly constitutes an APM often render its application null and void. For example, command-initiated directional mines are not victim operated and therefore fall outside the remit of the treaty;[2] meanwhile, the claymore mine (which often permits command-initiation) has become one of the most common weapons utilised by the Liberation Tigers of Tamil Eelam (LTTE), among others, along roads and paths in conflict zones, inflicting numerous civilian casualties.

Added to this ambiguity is the fact that the Government of Sri Lanka (GoSL) does not deploy APMs in an offensive manner thus giving it a moral superiority in relation to the Ottawa Treaty and enabling it to tie its participation in the Convention on Certain Conventional Weapons (CCW) to the peace process and ultimately to the LTTE's forsaking the APM,[3] an unthinkable move for a group that relies upon both conventional and guerrilla tactics. These issues are by no means particular to the conflict in Sri Lanka, which has not acceded to the Ottawa Treaty, but are rather more symptomatic of the quagmire of technicalities and minutiae likely to be encountered with any such sweeping humanitarian and human-security-centred international campaign which seeks concrete legal action against a similarly widely available weapon system.

The issue at stake with the implementation of the Ottawa Treaty is the abandonment of APMs—which indiscriminately kill, are difficult to completely remove and threaten local populations long after the conflict has ended or moved on. Implied in this widely popular humanitarian proposal is the necessity for a transformation in the way that most of the world's conflicts are conducted. The APM is a weapon of attrition and defence that is employed in scattered and unmarked areas; it is known as one of the most effective means to make an impact during a wider strategic stalemate. It is also a cheap and cost-effective force-multiplier for an otherwise outnumbered and/or outflanked force. In its most basic 'dumb' form it is potentially a weapon of wider long-term destabilisation, but with little immediate decisive result other than as a warning system. Unlike the banning and restriction of weapons of mass destruction, the ban on APMs imposed by the Ottawa Treaty was not initiated due to fear of any immediate devastating use; landmines are in use in all areas of the globe where there is call for them. Nor do they face the same stigma as attached to the WMD: there is no doomsday scenario in their continued use, but simply a lower quality in, and steady loss of, lives in populated areas where they are deployed. In this respect, landmines are usually dealt with more as an epidemic or natural disaster, with demining and mine-risk education the most effective responses, while the cause of their being deployed, however unpleasant, is usually treated as an almost immutable fact.

Intra-state war in Sri Lanka has conformed to a commonly expressed formula of territorial and ethnic violence which has produced an attritional and particularly vicious conflict that has been raging with varying degrees of intensity for more than two decades. In cases such as Sri Lanka, the Ottawa Treaty will provide little succour for those daily affected by the threat of landmines and it is precisely in these areas that humanitarian demining is most needed. Thus, while aid given to demining agencies within Sri Lanka assures a certain level of activity, the underlying objective of the Ottawa Treaty—to stigmatise the production, transport and use of the weapon to the extent that it is not purchased or used—will remain unattainable in this case in particular and in many conflict areas involving armed non-state actors in general, due to the nature of the fighting, motivation of the combatants and what is at stake for all involved. This is not to say that the ideal of banning the anti-personnel landmine is in any way flawed. The ideal is simply not relevant when facts on the ground translate into only cursory attempts to respect well established humanitarian protocols such as the Geneva conventions, let alone the altruistic motives of many of the signatories to the Ottawa Treaty.

2. The Effects of Landmine Use in Sri Lanka

While Sri Lanka is certainly not the most landmine-affected country, it does represent some of the more frustrating and difficult problems in combining demining with re-integration of displaced communities in a war-torn country. As one would expect, the most landmine infested areas of the island are those where the LTTE and Sri Lankan Army (SLA) have repeatedly skirmished and where villages have been captured and re-captured resulting in both defensive and nuisance minefields and, of course, varying degrees of unexploded ordnance (UXO) from both sides' artillery and the Sri Lankan Air Force (SLAF).[4] Despite surveys occasionally conducted by the GoSL and touched upon by the UN, exact estimates of contaminated land are impossible due to the varying intensity of operations conducted throughout the conflict, the peace negotiations and the recent failed CFA.

The Convention on the Prohibition of Anti-Personnel Mines faces its greatest challenges in cases such as Sri Lanka where the GoSL and LTTE both call on each other to cease using the weapon and will not sign the treaty until the other does so. Therefore an outright commitment by both sides to end the employment of the APM in the conflict is linked to negotiations and, ultimately, to a real commitment to end the fighting, something that as of the beginning of 2008 seems unlikely until either decisive victory is obtained or bloody stalemate resumed. Furthermore, since the 2002 CFA, mine clearance has become a factor in the highly political allocation of donor funds which aimed at building a peaceful solution through development projects in which mine clearance plays a crucial role.[5] As such, the moral issue of APM use, as well as the practical issue of their removal, continue to be intrinsically linked to the perpetual political struggle between the LTTE and the GoSL for public and international support.

On quite another level, demining (in areas that are outside the conflict area) provides an important example of using reconstruction to integrate local communities into the peace process where other tactics have blatantly failed.[6] Unfortunately, there appears to be little motivation in cultivating this, the most obvious aspect of international donor aid and support, into a wider strategy of re-building the infrastructure and the GoSL links with the north-east and bringing about meaningful negotiations with the LTTE.

Despite the apparent futility of continuing humanitarian demining during the ongoing conflict and the lack of real commitment by the LTTE to abandon the use of APMs as a nuisance weapon and in a mainly unmarked and scattered fashion, there has been a consistent effort on the part of various demining groups, in conjunction with the GoSL, to limit the effects of APMs on internally displaced persons (IDPs).[7] Lessons have certainly been learned and by 2007 there appears to have been a successful campaign of mine-risk education in combination with government resettlement programmes in the north-east which drastically cut APM casualties in the Batticaloa and Trincomalee districts for that year.[8] LTTE-controlled areas in the north have also benefited from an emergency humanitarian mine action strategy and it appears that once an area has been removed from conflict these various demining and mine education strategies, using SLA engineers as a vanguard, have a high rate of success in combining demining with IDP resettlement and education.[9] There is little information on how well re-integrated these communities have become or if they have remained so, although the low casualty figures reported are obviously a promising indicator of progress towards a return to some sort of normality for those most affected by the conflict. In this respect, Sri Lanka is indicative of many countries where lengthy civil wars and ethnic violence have meant that poor levels of security hinder any full scale effort at demining.[10]

Contrasted with the military-style demining operations after the Second World War, humanitarian campaigns, such as those now organised under the Ottawa Treaty, are simply influenced by too many factors, often wholly unrelated to purely disarming and removing APMs, to have the same level of efficiency. Demining, along with IDP re-integration and the infrastructural development necessary to allow the north-eastern area of the island a chance at rebuilding its economy, are all ultimately mitigated by the recurrent nature of the conflict. Thus, any mine-clearance strategy, however effective, will not eradicate civilian APM casualties as long as the conflict continues to be fought near towns, villages and farms, something seemingly inevitable given the nature of the conflict and the relatively small geographic limitations of the battlespace.

3. LTTE Strategy and the Use of Landmines

The employment of partisans, guerrillas, militias and other 'auxiliaries' has always been a facet of western warfare and has continually re-surfaced, albeit most often as an addition to the conventional main body of strategy. Early attempts to justify the intrusion by commoners and laymen upon what was, until well into the Napoleonic Wars (1792-1815), regarded as the business of gentlemen and afterwards of professionals, have largely focused on the predominately auxiliary role that these units have played during most of European history, although with a few notable exceptions and especially from the early modern period until the Cold War insurgencies and current LICs. The wide ranging impact of Revolutionary France and the levée en masse of 1793 irremediably transformed the conduct of warfare, both due to the drastically increased scale and, more importantly, the corresponding commitment now required of the participating state in terms of economic expenditure and the political and social re-ordering needed to raise the citizen armies necessary for these much larger wars.

A. The Transformation of Low Intensity Conflict in the Developing World

The subsequent conception of 'small war,' as discussed by Clausewitz,[11] while theoretically the starting point of guerrilla warfare, reflected the tactical niceties of light infantry with their disruptive and mobile abilities rather than a more sweeping insurgency against an occupying force. The 'Spanish Ulcer,' however, was a demonstration of the potential of this aspect of a people's involvement in warfare (despite the number of guerrillas themselves being relatively small) dominated by lines of march and long and ponderous lines of communication.[12] Clausewitz did, however, recognise a more general aspect of people's war as potentially separate from 'small wars,' where the will of an entire nation inevitably aided the defender once the enemy had reached his point of culmination. The industrialisation of warfare saw the role of the people change from bystander and occasional victim, to the main difference between the overall victory or defeat of nations. General Ludendorff recognised as much and was to demonstrate the importance of the will of the people and organised state propaganda in enabling the German Army to commit itself well above limits thought absolute before the First World War.[13]

However, Clausewitz' later realisation, expounded upon by Marx and Engels, that each particular 'epoch' had its own unique model of warfare provides a useful platform for examining the development of modern guerrilla warfare theory, its ramifications on westernised conceptions of war, its selective application by the LTTE and finally on the role of the APM weapon system in the Sri Lankan conflict. Implicit in the idea of revolution within a state for the purpose of either supplanting that state or, as in Sri Lanka, with the goal of creating a separate state within existing political boundaries, is the pre-requisite of the support and involvement of the people as a body.

While Clausewitz noted that the use of the people in war (in place of or in conjunction with the professional army) would indeed lend one side an advantage, Lenin codified the principle that oppression and injustice from above could, given the right leadership, result in a popular uprising where it was to be the people, transformed into a people's army, that achieved what were traditionally separate military goals. While noting cases where the state did not fully monopolise the use of force within its own territory, Lenin was careful to point out the fact that only properly structured revolutionary movements, led by the political elite, could use this force with any legitimacy and always in line with the interests of the working class, without being labelled as 'terrorists.' Legitimacy, therefore, came from the willingness of the people to be led, or at least to tacitly support a political elite, so that, in the final analysis, persuasion through propaganda and coercion was a pre-requisite for any act of guerrilla warfare that would facilitate a successful mass uprising.[14]

Additionally, Lenin noted the levels of violence and the radical form that the armed struggle should take precluded any outdated notions of honour in the conduct of warfare. While the First World War had demonstrated that there was little room left for such ideas in modern industrialised, total conflict, the inherent asymmetry in the people's war that Lenin discussed meant that this was much more the case.

The LIC strategy of the early Cold War period was codified by Mao Tse-Tung's writings describing the Chinese Communists' experience during the Japanese occupation and the following bloody civil war against the Kuomintang. As with Lenin, much of the focus of Mao's theory comes down to class division and ideology, although this was of less import during the initial survival stages of the insurgency. Here indoctrination is seen as being a facilitator of the widespread support, active or tacit, so essential for the continued survival and growth of the insurgency. It was this aspect that Mao termed 'sea,' an anonymous mass of supporting peasants in which the small number of guerrilla 'fish' could take refuge, making any attempts at striking at them appear heavy handed and directed against the people in general.[15] Added to this Mao recognised the careful balance inherent in guerrilla warfare, where poor morale and lack of political motivation on the part of the insurgents could easily isolate the population and expose the guerrillas to defeat in detail.[16] The will of the people, not only to actively resist their oppressors and fill the ranks of the insurgents, but simply to be prepared for the sacrifices involved in allowing the guerrillas to operate within their towns, villages and countryside, involves a delicate calculation that has underlined all popular uprisings and revolutions and inevitably calls for a degree of suffering amongst civilians that has become rare in the modern westernised way of war.

Furthermore, both Mao and Lenin fully recognised the principle described by Marx and Engels, that there is not one unified theory of guerrilla or revolutionary warfare and that the particular historical experience, culture, economy and political motivations of each case are of paramount consideration in any implementation of a strategy of guerrilla resistance, not to mention the nature and degree of violence adopted by the guerrillas. Finally, it is important to recognise that guerrilla warfare is not enough to achieve the end-state required by those conducting it. Mao fully recognises this fact and states that it is often required to revert to a previous 'stage' of resistance, often for prolonged periods of time, but always with the ultimate objective of fighting a conventional war in order to destroy the enemy's means of resistance and thus decide the political outcome which is the ultimate expression of the guerrilla movement.[17] Here it is possible to begin drawing parallels with the Tamil resistance in Sri Lanka and the background to the conflict which has dictated the levels of violence, the development of the methods of waging war and, ultimately, the place of landmine-type weapons within guerrilla warfare strategy and specifically, the use of these weapons as a factor in LTTE strategy throughout the conflict.

B. LTTE Strategy

Mass organised armed resistance by Tamil militant groups only began in earnest in the 1980s, by which time the most influential guerrilla conflicts of the twentieth century, both successful and disastrous, stood as invaluable examples of the power of armed popular resistance against colonial, imperialist and religiously intolerant governments wielding larger, more advanced and better organised armed force. Most importantly, recent guerrilla conflicts had demonstrated the longevity of this type of conflict; from the successful thirty year conflict in Indo-China/Vietnam to the bloody struggles in Mozambique and Angola and the continued stalemates in Korea and Cyprus, to name but a few.[18] The fact that the myriad Tamil resistance groups were predominately political and student organisations which were denied a peaceful avenue for pursuing their cause led to varied approaches to the initiation of guerrilla warfare, often borrowing heavily from the examples of Lenin, Mao and Che Guevara.[19] While all of these groups were nationalist in their outlook, only a few (most notably the powerful People's Liberation Organization of Tamil Eelam, PLOTE) were identifiably communist or Marxist. Furthermore, the contemporary examples of militarised political agitation set by Hezbollah and the Palestinian resistance were undoubtedly strong influences and were to be emulated, first by the Eelam Revolutionary Organization of Students (EROS) and later, with better success by the LTTE.[20]

The constellation of Tamil resistance organisations were condensed into the Tamil Five[21] during the late 1970s but the achievement of an independent state by force of arms alone gained little support due to the poor co-ordination and criminal appearance of these groups.[22] The deteriorating situation after the riots of 1983, however, led to an upsurge in Tamil support towards the militant groups who they saw increasingly as the only form of protection against a Sinhalese Army running rampant and a Sri Lankan government unwilling or simply unable to act to prevent them. This minimal support base enabled the LTTE to survive as a semi-legitimate nationalist group and begin the basic guerrilla operations of coercion, indoctrination and most importantly the elimination of political enemies.[23] While the LTTE was a far from palatable standard-bearer for Tamil independence, the GoSL effectively helped to create an environment which left this, the most ruthless of the Tamil militant groups, as the only choice for even the most moderate of Tamils. The GoSL's forcing of Tamils into the arms of the LTTE reached a culmination when the overarching Tamil party, Tamil United Liberation Front (TULF), was left no political recourse due to an amendment passed which effectively banned any political progress towards secession, the main goal of the TULF.[24]

Thus, building on a strong ethnic identity and exploiting the economic and political benefits provided by the Tamil diaspora, the LTTE was able to harness the political momentum and, although not strong enough to win a conventional war at this point, it, along with the other Tamil secessionist para-militaries, was able to create enough awareness and international concern for the regional government in Tamil Nadu, a long-time base of operations for most of the Tamil militants, to mobilise public opinion in India towards a settlement with the GoSL and a better position for Tamils in Sri Lanka, although the party which was the focal point for this, the AIADMK, stopped short of calling for full independence.[25] Crucially, it was the high levels of political violence, including early LTTE suicide bombings, that forced the Indian government to take notice. The desire of India's leaders to establish greater dominance over the region and its heightened military prestige in suppressing rebels in the north-east of India, combined with the very public outcry over perceived ethnic cleansing of Sri Lankan Tamils, led to a confused but powerful intervention in the conflict.[26] The net result of India's statecraft was to be the Indo-Lanka Accords of 1987 and the subsequent direct intervention of the Indian Peace Keeping Force (IPKF) which, in turn was to create a degree of unity of action between the LTTE and the GoSL—at least while the Janatha Vimukthi Peramuna (JVP) insurrection lasted—leading to increased suffering of Tamils in the north and east of the island as the IPKF-backed Tamil National Army (TNA), led by surviving cadres of TELO, ENDLF and EPRLF, sought to re-gain legitimacy for the Tamil cause while once again trying to avoid annihilation by their old LTTE adversaries.[27]

Throughout this period the LTTE conducted both conventional operations as well as countless guerrilla actions, always reverting to the latter as pressure from the more advanced and numerous IPKF and SLA units began to tip the balance of force in their favour at different periods of the fighting. Both the SLA and IPKF used landmines extensively, as defensive obstacles to be placed around bases and other strategic positions, in line with the Soviet development of landmine warfare. Tamil militants in general and the LTTE in particular, benefited greatly from extensive training given to them by Indian military advisors during a misguided attempt by India to exert its influence over the conflict. This training included communications, infantry tactics, intelligence gathering and important methods of operational command and control, all of which were employed by the LTTE, furthering their military capabilities and ensuring the survival of the organisation.[28]

The casualties sustained by the LTTE and, more importantly, their civilian support bases, led to renewed GoSL operations after the withdrawal of the IPKF which resulted in the capture of Jaffna in October 1995 and to a regression by the LTTE to a more low-intensity campaign using the tactics of guerrillas and insurgents and focusing on high level political targets rather than capturing land and inflicting military defeats along the front lines.[29] Evidence of this regression can be found in the re-armament and re-organization undertaken by the Tigers during the initial period of the Ceasefire Agreement of February 2002 and in their rejection of a moderate political solution demonstrated by their enforcing a boycott in the north of the country of the 2005 elections, essentially gifting Mahinda Rajapakse the presidency at the expense of the moderate candidate, Ranil Wickremasinghe, with the expectation that this, combined with an increased campaign of political terrorism, would restart the conflict proper with the Tigers gaining the moral high ground.[30]

The LTTE made a strategic blunder in mid-2006, however, by attempting to goad the GoSL into an all-out conflict by cutting the water supply from the area under its control into more ethnically mixed areas which resulted in an unexpectedly powerful response from the SLA and security forces which, in turn, led to the capture of Sampur in the Trincomalee district and the rout of LTTE cadres in the east. Further SLA operations resulted in some twenty training and operational camps being captured by government forces. Strategically much of the Tigers hard-won ground in the east had been lost, often due to the employment of large numbers of conscripts and the overall effect of the 2004 Tsunami on the LTTE infrastructure, not to mention the defection of Colonel Karuna along with some 5,000 eastern cadres, traditionally the most elite of the regular troops, from the LTTE, resulting in a significant drop in the quality of the troops available.[31] These losses have put a premium on weapons of destabilisation, namely landmines (mainly the claymore) and suicide attacks. With the formal end to the CFA the GoSL has resumed large scale military operations against the LTTE, seeking to build upon their success in the east. As of March 2008 the conflict appears to have resumed the stalemate that has characterised operations for much of the war with the Tigers falling back on guerrilla strategy as government pressure begins to gain territory.

4. Civil War in Sri Lanka within the Context of the Ottawa Treaty

The use of landmines in the conflict in Sri Lanka is not solely responsible for the disruption and misery that has engulfed the northern and eastern areas of the island and, more sporadically, the capital city and many other Sinhalese towns in the centre and south. This is not to say that the effects of the widespread deployment of this weapon are meaningless, simply that the LTTE have adopted this weapon as a matter of course and that, despite its inherently controversial nature, the landmine is certainly not the most unethical or immoral means of killing employed during this conflict.

That the APM is indiscriminate is beyond doubt; however, the ravages of the IPKF in the late 1980s and early 1990s, the paramilitary killings and those committed by the SLA and other Sri Lankan security forces were hardly guided by any meaningful intelligence information. In the same vein, the seemingly tactical and occasionally strategic use of Black Tiger suicide bombers by the LTTE against political and military targets do not discriminate against bystanders just as a precision guided cruise missile does not. The main aspects that set the APM, as defined by the Ottawa Treaty, apart from any number of wholly indiscriminate weapon systems are that it is entirely victim operated, thus the victim theoretically is in charge of its own demise or injury as its force placed on the mine detonates it and it is deadly long after the battle space has been vacated and returned to a state of peace.

The APM is therefore labelled as being an unnecessarily perfidious weapon and the stigma against its use is based upon the unnecessary death and injury of civilians, as well as the associated effects on the local economy and communities. The moral crusade to ban the landmine can thus be viewed as a noble and just endeavour. However, seeking to ban the use of landmines in a state where a conflict rages that has shown all the hallmarks of ethnic and civil violence and where the use of suicide attacks have proliferated is an unfortunately irrelevant cause when the moral limits of the warring parties have, as in most other civil wars and insurgencies, been set far beyond those of most of the signatories to the Ottawa Treaty.

A. LTTE Use of Suicide Attacks as a Strategic Weapon

Perhaps the most important contributor to western perception of the LTTE is their extensive use of suicide units, called the Black Tigers, as a strategic and often tactical battlefield weapon. In purely military terms the LTTE have effectively created a precision guided weapon which is often initiated by the most obvious and primary victim, the suicide bomber. The willingness of individual Tamils to kill themselves, or place themselves in situations where death is all but inevitable, in order to achieve what are often politically motivated and well calculated assassinations is particularly galling to many westerners. Originally sympathetic to the oppression of an ethnic minority group and their struggle for recognition on the world stage, begun in earnest in the late 1970s, western perceptions have been altered by the escalated use of violence and the moral taboos broken by the unpleasant and vicious conduct of the war as epitomised in the Black Tigers.

Although ideology and oppression are motivating factors, aspects of Tamil culture and traditions are reflected in the manner in which the Black Tigers are represented to the wider Tamil people. The suffering and ingrained psychological trauma of Tamil communities within Sri Lanka allow for the justification, and even reverence, of suicide bombers or attackers and we therefore assert that within this community any moral justification for the cessation of the use of landmines is bound to be blunted by an already psychologically over-laden population. Furthermore, the deleterious and unique effects of the landmine, in that it is a psychological target-operated weapon, while insidious and damaging to the social fabric of rural communities are only one source of the extreme and often random violence experienced by rural communities from any number of sources. While the SLA and indeed the IPKF, have been known to attack civilian population areas with artillery and air power with highly destabilising effects on these communities and of course leaving UXO and air scattered landmines in the areas, it is the LTTE high profile use of suicide bombers as well as their apparently deliberate targeting of vehicles with various mines and improvised explosive devices (IEDs) with little regard for civilian casualties that has placed the Tamil Tigers on the same list as Al-Qaeda and Hezbollah.

B. Rationalising the Use of Terror in the Sri Lankan Conflict

All sides during the two-decade old conflict have been guilty of using what are regarded as 'terror tactics' at some point. The two main protagonists have varied in their employment of terrorism depending upon the international climate, success or failure on the battlefield, outsider influence and the presence of domestic rivals on either side's cause. The LTTE launched its unprecedented campaign of coercion, public suicide bombings and assassinations in the mid-1980s in order to solidify its position as the only legitimate Tamil militant group.[32] On the other side of the spectrum, the GoSL is guilty of having employed state terror in its initial violent reaction to Tamil civil disobedience in the late 1970s with the Prevention of Terrorism Act giving police and security forces unprecedented powers and was also guilty of, at the very least, turning a blind eye to the suspiciously well organised looting, rape and murder committed by enraged Sinhalese during the political watershed of the 1983 riots.[33]

It is perhaps inevitable that political violence on an ethnic level has led to civil strife, which has given rise to armed conflict, which has in turn deprived the successive governments in Sri Lanka of their monopoly on the use of force. What is certain is that after so many years of violence, not only from the LTTE and GoSL, but also from their partners, off-shoots and domestic rivals (the Karuna faction and the JVP to name but two), consideration of the laws of war and those governing human rights are now only honoured in the most superficial and disingenuous manner. It is within this context that the use of terror tactics and misinformation concerning ongoing human rights violations and government and LTTE oppression must be viewed. The utility of the landmine and IED as a weapon system deployed by both sides in the Sri Lankan conflict needs to be placed within this much wider perspective.

The seemingly indiscriminate employment of APMs by the LTTE and the specific targeting, often of civilian transport, by command initiated devices such as the claymore mine and IEDs, factors greatly in both domestic and international perception of the organisation. There is, however, nothing unique in the use of landmines and other 'trap' weapons in the Sri Lankan conflict. Underlying the employment of these weapons is the innate inferiority of the LTTE and other militant or paramilitary groups, in terms of numbers, technological capability and material, which has necessitated the adoption of relevant aspects of guerrilla warfare theory which has been developed and refined during the twentieth century. Revolutions, civil wars and wars of national liberation from colonial or imperialist forces have all drawn on the role of the civilian in the conduct of war as a source of strength and indeed of survival.[34] As such, the traditional (western) value-system, which has defined and attempted to limit the unpleasant extremes of warfare throughout modern history, has had great difficulty in dealing with the proliferation of the low intensity conflict (LIC).[35]

At the heart of the ban on the sale and proliferation of landmines and perhaps that of small arms in the future, is the assumption that now that the era of total and theoretically absolute war appears to be at an end, it is possible to re-instate some limit, not on the conduct of war, but on the very weapons with which it is fought. While this is undoubtedly a noble goal, a number of eternal principles, or undying truths, of warfare prevent many of those who participate in LIC from accepting the tactical, sometimes even strategic, disadvantages resultant from the banning of any weapon that they possess and especially one as versatile, cheap and numerous as the landmine. Here moral arguments also fall short when one side or both are willing to use what amounts to suicide tactics which implicitly accept yet another form of indiscriminate violence, although it is often command initiated, if one views the suicide attacker as a rational military tool.

5. Conclusions

Contemporary intra-state conflicts are, to varying degrees, outside the remit of international law as, short of lengthy economic sanction or direct military intervention, there is simply no realistic enforcement mechanism in place. Thus it should be of no surprise that the Ottawa Treaty is of little consequence in countries and regions ravaged by these conflicts; the APM is either too valuable a weapon system or there is too little incentive for guerrilla forces to cease using them while they are facing more advanced, better equipped and often more numerous foes. Furthermore the nature of the violence in many of these conflicts, especially those involving institutionalised use of suicide attacks, complicates the moral arguments for banning the APM due to its being an indiscriminate weapon system. The suicide bomber used by the LTTE as a tactical weapon inevitably strikes in public leaving unintended victims maimed and dead. Similarly the command detonated IED or claymore mine, while not victim-initiated, is inevitably indiscriminate within its arc of destructiveness. Again, the same can be said of artillery and air strikes, all of which can leave UXO which often have the same disastrous effects as scattered, unmarked minefields.

While it is true that some form of international legal action is better than none at all, even as a simple statement that this underhand form of warfare is not acceptable to a moral standard set forth by the signatory nations, the simple fact is that landmine warfare will continue in some form or other whether or not the guerrilla forces have access to Pakistani or American built mines, or whether they are hastily improvised from domestically available material. This being the case, the efforts of humanitarian and government demining agencies and units combined with a well thought-out campaign of mine-risk education for those most in danger of

APMs and UXO is often the most effective use of resources as it is not likely that the immediate future will see adhesion to the Mine Ban Treaty by some of the main exporters of APMs to the developing world.

Human-security-based campaigns and treaties calling for the abolishment of the weapon systems with which modern insurgencies and civil wars are fought, namely the APM, IED, AK-47 and RPG-7 (or in several extreme cases the machete), must in reality seek to tackle the issue of state failure, ethnic or religious differences and nationalist movements from the bottom up. Unfortunately, the systemic issues at stake that are unable to be solved by diplomacy or international coercion are unlikely to be brought to peaceful conclusion, or even noticeably restrained, by proscribing certain weapons seen as being too indiscriminate.

Furthermore, once one weapon is labelled as illegal or immoral, there is a strong moral argument to then widen this claim. Indeed, the unintentionally callous technicalities behind defining prohibited landmines exclude other weapons that often have the same, or similar, effect; UXO is a prime example of this. There is an argument that the overall aim of the Ottawa Treaty and the ICBL is based upon a much more long-term view, that imposing some form of international stigma on certain weapons will, eventually, impact their use in states not under the jurisdiction of the Ottawa Treaty. Unfortunately, many non-state actors who frequently employ APMs are simply not susceptible to international, largely western, condemnation as they are usually already condemned by these selfsame parties, as is the case with the LTTE.

What is more, certain states that have signed the treaty have already demonstrated that, although they have blatantly violated it for whatever ends, there is little in the way of punishment for their actions, as the threat of these often unstable states leaving the treaty is seen as being potentially more damaging than their continued, though disingenuous, participation. Thus, the Ottawa Treaty has helped create what amounts to a zero-sum environment where one either prohibits APMs completely, or is indirectly classified as being a party to their continued use throughout the Developing World. One of the primary reasons for many states failing to sign up to such a human security based treaty is, unsurprisingly, their unwillingness to part with an entire weapon system which, although not decisive, has proven to be integral in the conduct of many conventional and low intensity conflicts throughout recent military history.

The treaty therefore fails to account for the realist, state-level notions of security so vital for Third World governments who are fighting insurgencies and whose military balance of power remains on a knife-edge. While not the most landmine-affected country, Sri Lanka is representative of many of the problems, both practical and ethical, with the Ottawa Treaty, which seeks to outlaw a weapon on the basis of a moral justification almost entirely western in origin, that has often become a way of war in itself and which has become so dominant in conflicts in the developing world.

Notes

[1] M. Croll, *The History of Landmines*, Leo Cooper, Barnsley, 1998, p. 128.
[2] 'Landmine Monitor 2007 Report: Sri Lanka,' Accessed February 7, 2008, <http://www.icbl.org/lm/2007/sri_lanka.html>.
[3] Ibid.
[4] Ibid.
[5] C. Orjuela 'Reaping the Harvest of Peace?' *Critical Asian Studies*, Vol. 40, No. 2, 2008, p. 216.
[6] Ibid, p. 217.
[7] 'Mine Action's Mid-Year Review of the Portfolio of Mine Action Projects, 2007,' Accessed February 5, 2008, <http://www.mineaction.org/downloads/1/MYR%20Final.pdf>.
[8] Ibid.
[9] Ibid.
[10] Croll, op. cit., p. 129.
[11] B. Heuser, *Reading Clausewitz*, Pimlico, London, 2002, p. 134.
[12] S. Kinross, 'Clausewitz and Low-Intensity Conflict,' *The Journal of Strategic Studies*, Vol. 27, No. 1, March 2004, p. 37.
[13] H. Speier, 'Ludendorff: The German Concept of Total War,' *Makers of Modern Strategy: Military Thought from Machiavelli to Hitler*, E.M. Earle (ed), Princeton University Press, Princeton, 1973, p. 315.
[14] V.I. Lenin, *Lenin Collected Works*, Vol. 11, Progress Publishers, Moscow, 1965, p. 216.
[15] R. Smith, *The Utility of Force: The Art of War in the Modern World*, Penguin Books, London, 2005, p. 169.
[16] M. Tse-Tung, *On Guerrilla Warfare*, trans. S.B. Griffith, University of Illinois Press, Chicago, 2000, p. 92.
[17] Smith, op. cit., p. 169.
[18] Ibid., p. 289.
[19] B. Mathews, 'Radical Conflict and the Rationalization of Violence in Sri Lanka,' *Pacific Affairs*, Vol. 9, No. 1, December 1985, p. 34.
[20] N.M.R. Swamy, *Tigers of Lanka: From Boys to Guerrillas*, Konark Publishers PVT LTD, Delhi, 1995, p. 100.
[21] Tamil Five consisted of: TELO, EROS, PLOTE, EPRLF and TNT (later the LTTE).
[22] Mathews, op. cit., p. 36.
[23] M. Bloom, *Dying to Kill: the Allure of Suicide Terror*, Columbia University Press, New York, 2007, pp. 53-54.

[24] Ibid., p. 54.
[25] Ibid.
[26] J.P. Senaratne, *Political Violence in Sri Lanka, 1977-1990: Riots, Insurrections, Counterinsurgencies, Foreign Intervention*, VU University Press, Amsterdam, 1997, p. 96.
[27] Ibid., p. 100.
[28] Ibid., p. 96.
[29] Hopgood, 'Tamil Tigers 1987-2002', *Making Sense of Suicide Missions*, D. Gambetta (ed), Oxford University Press, Oxford, 2005, p. 49.
[30] ISS Strategic Comments—Conflict in Sri Lanka, p. 1.
[31] C. Smith, 'Sri Lanka Returns to War,' *Asian Affairs*, Vol. 39, No. 1, March 2008.
[32] R.A. Pape, *Dying to Win: The Strategic Logic of Suicide Terrorism*, Random House Publishing, New York, 2005, p. 141.
[33] Bloom, op. cit., p. 53.
[34] Kinross, op. cit., p. 38.
[35] J. Keegan, *A History of Warfare,* Pimlico, London, 1994, p. 3.

Bibliography

Battin, M.P., 'The Ethics of Self-Sacrifice: What's Wrong with Suicide Bombing?' *Archives of Suicide Research*. Vol. 8, No. 1, January 2004, pp. 29-36.

Bloom, M., *Dying to Kill: The Allure of Suicide Terror*. Columbia University Press, New York, 2007.

Croll, M., *The History of Landmines*. Leo Cooper, Barnsley, 1998.

Heuser, B., *Reading Clausewitz*. Pimlico, London, 2002.

Hopgood, 'Tamil Tigers 1987-2002.' *Making Sense of Suicide Missions*. Gambetta, D. (ed), Oxford University Press, Oxford, 2005.

Keegan, J., *A History of Warfare*. Pimlico, London, 1994.

Kinross, S., 'Clausewitz and Low-Intensity Conflict.' *The Journal of Strategic Studies*. Vol. 27, No. 1, March 2004, pp. 35-58.

Lenin, V.I., *Lenin Collected Works, Volume 11*. Progress Publishers, Moscow, 1965.

Mathews, B., 'Radical Conflict and the Rationalization of Violence in Sri Lanka.' *Pacific Affairs*. Vol. 9, No. 1, December 1985, pp. 28-44.

Orjuela, C., 'Reaping the Harvest of Peace?' *Critical Asian Studies*. Vol. 40, No. 2, 2008, pp. 211-232.

Pape, R.A., *Dying to Win: The Strategic Logic of Suicide Terrorism.* Random House Publishing, New York, 2005.

Senaratne, J.P., *Political Violence in Sri Lanka, 1977-1990: Riots, Insurrections, Counterinsurgencies, Foreign Intervention.* VU University Press, Amsterdam, 1997.

Smith C., 'The Eelam Endgame.' *International Affairs*. Vol. 83, No. 1, 2007, pp. 69-86.

—, 'Sri Lanka Returns to War.' *Asian Affairs*. Vol. 39, No. 1, March 2008, pp. 83-94.

Smith, R., *The Utility of Force: The Art of War in the Modern World.* Penguin Books, London, 2005.

Speier, H., 'Ludendorff: The German Concept of Total War.' *Makers of Modern Strategy: Military Thought from Machiavelli to Hitler.* Earle, E.M. (ed), Princeton University Press, Princeton, 1973.

Swamy, N.M.R., *Tigers of Lanka: From Boys to Guerrillas.* Konark Publishers PVT LTD, Delhi, 1995.

Tse-Tung, M., *On Guerrilla Warfare.* Trans. Griffith, S.B., University of Illinois Press, Chicago, 2000.

Graeme Goldsworthy is one of the foremost international experts on Small Arms, Light Weapons, Landmines and the Explosives Remnants of War (ERW), having been an officer in the British Army. During and after his service he managed landmine and UXO clearance operations in South Asia, Africa and Central America for a number of international organisations. He is currently a recognised expert on International Security for the Hague Centre for Strategic Studies, leader of the Harvard (ICDR)—VUMC study, focusing on the impact of landmines and other SALW (Small Arms / Light Weapons) in conflict, in international law and in terms of psycho-social impact on vulnerable populations. Graeme is also an adjunct lecturer at Webster University, visiting lecturer at Rhodes University, South Africa and a number of other institutions in the United States, the Netherlands and the United Kingdom.

Toby Chesson is a content researcher and provides training support for the Pax Ludens Foundation in The Hague, The Netherlands. He is a graduate of King's College London with a B.A. in War Studies, and an M.A. in the History of Warfare in which he focused upon low intensity conflict, particularly in Sri Lanka.

Erica Pasini holds an MSc in Pharmaceutical Chemistry, an MA in Journalism and a Ph.D. in Malaria Drug Development Research from the University of Milan, Italy. She is and has been a post-doctoral research fellow in the field of malaria at several leading EU institutions. During the last few years, Dr. Pasini became interested in the geopolitical, legal and health related aspects of complex emergencies and successfully completed an MA in International Relations. At present, she is actively pursuing research and field experience in the human security realm concentrating specifically on the Middle East, the Balkans and South East Asia.

From Manifest Degradation to Latent Anticipation: Military Boredom in the First World War and Afghanistan

Bård Mæland

Abstract
Scholarly articles as well as headlines from news reports indicate that the phenomenon of boredom is an intimate part of the kind of operations we see in our time. Yet, boredom in a military environment is hardly a novelty. Imagine the times of waiting in the US Civil War, the life in the trenches of WWI, or marches deep in the jungle of Vietnam. If this is the case in conventional war, it surely applies to a number of current military operations and actions. In these, soldiers may be exposed to a sense of boredom that may well turn out to be a serious and enduring challenge, often regarded as an enemy to be fought in the midst of oneself. Consequently, given that boredom may appear as a most serious threat for military units serving in diverse kinds of current operations, how may this state of mind be interpreted adequately as a human phenomenon and how may soldiers be trained to cope with it effectively? These questions are discussed in the light of historical material, especially from the Great War, philosophical approaches to boredom and an in-depth study of Norwegian soldiers who recently served in Afghanistan.

Key Words: Boredom, military, First World War, Richard Aldington, George Winterbourne, peacekeeping, Afghanistan, Norwegian.

*****

Wars, pogroms and persecutions have all
been part of the flight from boredom...
Boredom is a vital problem for the moralist,
since half of the sins of mankind are caused by it.
Bertrand Russell

My despair is less despair than boredom and loneliness
Anthony Swafford - *Jarhead*[1]

1. Military Boredom as a Research Topic

Smith & Wesson (revolver) was the father.
Boredom was the mother.
What kind of offspring you
would expect from such parents?

But that's how the game was born:
the man gambled with death
and his life was his only stake.[2]

The quotation is taken from an Internet article on how Russian Roulette was invented in the trenches during the Russian-Turkish Wars in the late 19th century. The article claims that it resulted from a combination of acute boredom and military inactivity, often 'aggravated by unreasonable consumption of vodka.' Boredom may of course express itself in less desperate forms. It seems, however, to have accompanied military life and military operations throughout history.

Lebanon late October 2006: The German *Spiegel Online* reports, 'The greatest danger for the soldiers is boredom,' after having visited Spanish elite soldiers deployed along the Blue Line between Southern Lebanon and Israel. After the Israeli war against Hezbollah in July of the same year, the international community had requested a robust force to police the border between Northern Israel and Southern Lebanon. But there was little to do for the soldiers. 'We are here just because we are here,' said one of the soldiers interviewed. A lack of motivating reasons, often combined with high military skills, is probably the perfect recipe for military boredom. But a seemingly boring situation can change swiftly. In late June 2007, almost a year later, three Spanish and three Colombian soldiers were killed in a bomb attack in the same area.

Some twenty years earlier, a rare study on boredom in peacekeeping operations commenced by stating that 'although boredom is one of the major problems experienced by those engaged in peacekeeping missions, there is little consensus on the nature of the problem.'[3] The study examined boredom and associated phenomena in the first American battalion to serve in the Sinai Multinational Force and Observers (MFO) during the early 1980s. But, as the study implied, there is a need to create a deeper understanding of the complex phenomenon of boredom in a military context. Not only since it is an interesting phenomenon to delve into, but because it is important and vital for any military involvement to have an informed understanding of it. It is therefore interesting to find that boredom in 1992 was made an explicit theme in a field manual issued by the US Army for service in low-intensity conflicts.[4]

These recent examples from peacekeeping missions aside, military boredom is hardly a recent phenomenon in the history of humankind. Boredom has been a much-felt problem at all times. Interestingly, we can see traces of this in another contribution within this volume. In the chapter by Mustafa Serdar Palabıyık on the developments of the perception of war within the Ottoman Empire, boredom seems to be implicit in what is described as the distress and depression of the long wars of the 16th and 17th

centuries. As Palabıyık mentions, the mood of Ottoman individuals was that of longing for peace and being bored by wars. On the other hand, war also seems to have worked as a 'tool of motivation.' Sometimes war was admired in its peculiar violence and grandeur. One example here is the letters of Enver Paşa (1881-1922), who at the eve of the War of Tripoli and the Balkan Wars, in his letters to an unidentified young woman he might have met while serving in Germany, expressed the attraction of war to a young and ambitious soldier. So, even though wars may turn out as deadly boring, they may in the first place convey a certain attractiveness.

A standard formula says that wars consist of '5% horror and 95% boredom/waiting.' Imagine the life in the trenches in World War I, the jungle marches in Vietnam, deployment at Diego Garcia, how the development of media (radio, shows, TV, PC-games) throughout the 20th century helped soldiers to combat (!) boredom. Despite war being considered a situation of continuous fight and combat, countless stories told of in letters and biographies from the battlefields unveil another trait of war: being bored by waiting, lack of action and activity, but also a lack of concern and attention from the public back home—all of this immediately implying a lack of meaning. In this sense, military boredom is *historically enduring,* but it is also enduring in terms of what it demands and requires of individual soldiers who have *to endure it.*

No wonder that all kinds of specialists in and outside the armed forces of most countries have been called upon to cure it: welfare personnel, chaplains, soldiers' unions, commanders and parents' organisations. Currently, a big market of packages for 'boredom buster' has emerged, so called 'military care packages' that can be purchased online and sent to US soldiers suffering from lowered morale and boredom in Iraq and elsewhere.[5] In our age, with troops increasingly being required to go on deployments overseas, it may very well be that the service of an average soldier has become more meaningful, exciting and rewarding in many respects. But one important concern has to be addressed by those who send young men and women to conflict areas worldwide—that is, the 'over-stretched' use of soldiers overseas, with repeated postponements of their dates of return.

Yet, how serious is boredom, actually? Generally speaking, boredom can range from being a rather simple and superficial irritant to a grave, disease-like state of mind. Seán Desmondo Healy calls the former 'simple boredom' (*ennui commune/banal*), whereas the latter is called 'hyperboredom.' The former he defines as 'feelings of tedium that are at least in principle conscious responses to specific irritants that tend to goad one into escape.' The latter is understood as 'a deep-seated agony, scarcely realized except by its effects, which is brought on by an all-inclusive, persisting perception of what is taken to be one's existential situation.'[6] Consequently, boredom is a relative phenomenon, which in a military context is unlikely to be experienced in the same way by people who serve in the same unit, in the same place and at the same time. Boredom can serve as a precondition for imaginative and creative processes, but it can also form a devastating force of dissolution and destruction. In the most serious cases the effects of it can be quite devastating, as with how US military psychiatrists have counted boredom among the conditions that fuelled the harassment against prisoners in Iraq and Afghanistan.[7] This resembles the understanding of boredom by Søren Kierkegaard in *Either-Or*, where boredom is perceived as a quiet phenomenon, yet with an immensely destructive potential.

It remains, as mentioned, to examine the role of boredom with respect to military training and military life, or wars and other kinds of military operations. This will be done in two different ways in the following: first, I will present how boredom was experienced by one of the Great War soldiers, Richard Aldington, based on his famous book *The Death of a Hero* about the British soldier George Winterbourne (1929). This will show boredom in its rather comprehensive and exasperate mode. Then, I will present findings from a qualitative study I conducted about the experiences of military boredom among Norwegian infantry soldiers in Afghanistan during the present decade, where we find a much lighter form of boredom. In both cases, however, boredom has been examined as an explicit (i.e., mentioned as such by the informant) phenomenon, as well as an implicit (presupposed) one.[8]

2. The Great War Boredom of George Winterbourne

In 1929, after a decade or so of general literary silence, the best-seller *Death of a Hero* by Richard Aldington was published. Through the main character, George Winterbourne, a pre-war artist in London, the book has clear parallels to the author's own life. The effect of writing the book remained important to Aldington. From his autobiography one learns that he 'purged [his] bosom of perilous stuff that had been poisoning [him] for a decade.'[9]

The first part of the book takes place in London, where George is making a life as an artist in a rather bohemian environment marked by contempt for the values of Victorian society, especially its sexual hypocrisy and repression. Through his relationship with two ladies, Fanny and Elizabeth, George pursues sexual liberation and love of life, despite 'his rages and his worryings and the complications and boredoms.'[10] On the outbreak of war, he only hesitantly enlists in the army. He does not believe in the alleged causes of the war, which he does not find at all convincing. On the other hand, his political ideals mean that he wants to share the burdens of the soldiers in the ranks. And it was indeed a burden. It was a process of degeneration through which he not only lost his past and his present, but also his future—literally.

Describing his time in the army in retrospect, he mentions in one place a threefold strain, namely: his personal life, the strain of 'exasperation with Army routine,' and that of battle.[11] This threefold perspective may be useful for exploring the role of boredom in the literary biography of this First World War soldier. We shall begin, however, with the strain of 'Army routine,' then move on to battle strain and end up with the strain on his personal life, especially how the war removed him from his women back in London.

The upcoming war was much welcomed by George. Though he was doubtful about its causes on the British side and despite all the parading and tedious routines in his first training period, he saw the war as a relief. The war represented something real and it offered close contact with genuine masculinity. We may here see more clearly expressed what Palabıyık only hints at in his chapter on the Ottoman views of war elsewhere in this volume. Moreover, George joined the masses in their enthusiasm over the outbreak of the war. Interestingly, despite seeing the war as realising his masculinity, the 'pleasures' of the Great War seem to have also attracted female soldiers, as demonstrated by Joanna Bourke in her *An Intimate History of Killing.*[12]

However, boredom would rapidly dampen his hopes and expectations. His travel to the front was marked by monotony as he was introduced to the 'curious fact' of waiting, the lack of solitude and inaction.[13] Moreover, the weather and nature—the aesthetics of war (the greyness of heavy rain, the coldness of night and winter, the mud of the spring thaw, the smell of the trenches)—would also soon add to this feeling of comprehensive and intolerable boredom. After a while, George reached the first of two different 'expressionless' stages: 'He had reached the first expressionless stage of the war soldier, which is followed by the period of acute strain; and that in turn gives place to the second expressionless stage—which is pretty hopeless.'[14] The war would soon pull him into a process of 'interminable holocausts and immeasurable degradation of mankind.'[15]

In this situation, he suffered mentally, he felt that he had lost emotion and recognised how degeneration had set in. He became indifferent to his past (the ladies!), his ambitions faded and the repetitive mindset of 'depression, monotony, boredom' describes his situation precisely. It was a 'peculiar sort, a strained, worried, exasperated sort.'[16]

The notion of time is reflected in George's self-reflection. Time, experienced as quantitatively prolonged ('day' was experienced as 'month'), is likened to torture.[17] Previously, time had been a vehicle of joy and expectation, but once in the trenches, 'the hour-goddesses who had danced along so gaily before and have fled from us since with such mocking swiftness, then paced by in a slow, monotonous file as if intolerably burdened.'[18] At this time, the enemy has not yet put his company under serious attack. There is only 'mild artillery fire, sniping and machine-gunning,' and he starts thinking that 'the dangers of the war had been exaggerated, while its physical discomforts and tedium had been greatly underestimated.'[19] Thus, boredom had not only to be dealt with, it even forced itself upon those who tried to ignore it.

In this perspective, to go to the latrine, to get an easier position in the company, to see rats for the first time, to be exposed to dangers or even to be wounded would all be seen as attractive, simply because it would give you a chance for some diversion. Letters and parcels were also popular, nurturing a sense of home and feelings of sympathy. The troublesome insight of all this, according to Winterbourne, is that men have feelings. And, since replacement by robots was not an option, determination and endurance were required. This also applied to military success itself: the side that could destroy the other side's 'man-power and moral endurance' would win.[20] Boredom thus reveals the human condition in all its facets: the burden of boredom, the need for diversion and the potential for degradation.

However, through this jungle of depression, monotony and boredom, he still kept his 'consuming curiosity to see what the war was really like.'[21] He still hoped for movement, something real, relief and alleviation from the pain he felt as he waited to get to the frontline. Once on active duty, having left training in England, he felt an 'immense relief' as he would now face 'dread realities, not Regular Army pedants and bullies.'[22]

His wish to take part in the war was reinforced by his encounter with experienced soldiers coming back from the front. These men were men, 'strangely worn and mature, but filled with energy, a kind of slow, enduring energy.'[23] In comparison, the soldiers about to enter the war zone seemed both childish and feminine. The war offered an almost exclusively masculine experience, 'where no woman and no half-man had ever been, could endure to be.'[24] Soon he would be one of them. He was 'almost happy' and thought the war had 'saved something from a gigantic wreck,' namely 'manhood and comradeship, their essential integrity as men, their essential brotherhood as men.'[25] This description is ambivalent, however. Yes, the war offered an attractive masculinity to people bored of civilian life back in England, but what seemed attractive was also deceptive and destructive because it would soon ruin millions of male bodies. One could argue here that, despite the duty to go to war and the 'impiety' to demur,[26] boredom—i.e., the desire for active service—indirectly brought degeneration to the men who were first relieved and then destroyed them both physically and mentally.

One important expression of the latter is how soldiers understood themselves as numbers in the war. It soon seemed to Winterbourne 'insane to think that you had any individual importance.'[27] But nihilism went deeper as deterioration became physical and 'his body became worthless.'[28] Hence, he asks desperately: 'To be violently cast back into nothing—for what? My God! for what? Is there nothing but despair and death? Is life vain, beauty vain, hope vain, happiness vain?' His questions resemble the tone of the biblical Ecclesiastes: 'all is vanity.' This is existential boredom, which would soon turn into physical nihilism. 'He was surprised and a little alarmed at his own listlessness and despair. He felt like a sheet of paper dropping in the jerks and waverings through grey air into an abyss.'[29] This is nihilistic boredom, the boredom of meaninglessness, which is central to any boredom, yet in highly different ways.

Fear soon became the other big challenge for George, not to say the real enemy. During the first part of George's first visit (of two) to France one could say that fear and boredom were separated as night was separated from day. Whereas the night was the time for nervous waiting ('stand-to') for the German shelling and shooting, not least in the early morning hours, boredom, monotony and physical strain marked the days. This division of emotions and states of mind is well captured in *In Parenthesis,* David Jones's sketch of Private John Ball and how this soldier experienced the difference between daytime and nighttime in the trenches of the wasteland before the battle of Somme:

> Slowly they made progress along the traverses, more easy to negotiate by light of day. Not night-bred fear, nor dark mystification nor lurking unseen snares any longer harassed them, but instead, a penetrating tedium, a boredom that leadened and oppressed, making the spirit quail and tire, took hold of them, as they went to their first fatigue.[30]

This description would also have neatly fit Winterbourne's situation. Later during this period, however, fear takes over completely. There is simply no more room for boredom, only the continuous struggle to control one's nerves through the hailstorm of bullets and shrapnel bursts. At this stage of the war, it became 'absurd to talk about men being brave or cowards,'[31] and what mattered, despite individual differences in sensibility, was conscious self-control. As the constant pressure on one's nerves kept increasing, all one's energy and determination had to be repressed. One simply became thoroughly exhausted over time as the 'real' and 'intensified' nightmare commenced:[32] mud, systematic gas bombardment, incessant artillery, a steady rumble and roar and sleep deprivation. Winterbourne was of a sensitive nature and thus developed neurosis quickly. His war soon turned into a 'timeless confusion, a chaos of noise, fatigue, anxiety and horror.'[33] Perhaps this may even amount to a kind of boredom, though of a very different kind to that of monotony and routine?

It is a man exposed to such circumstances, 'a little mad,' with a 'cut in his life and personality,' marked by apathy, who returns to London after seven months to join the Officer Cadet Corps, before again going back to

France to end his life as an officer. As he arrives back at Victoria Station from his first tour to France, Elizabeth is there to welcome him, but she does not recognise him right away. He has changed. The next day, he is invited to see Elizabeth's friends, including her new lover. He behaves like a stranger to them. And they all seem strange to him. He feels he does not fit in any more. After the party, he returns to Elizabeth's apartment, where he lives and, in Elizabeth's room, he finds a draft of a letter to her lover dated the same night as he returned, starting thus: 'DARLING,—What a bore, as you say!'[34] Also when he later talks to Fanny, he gets the impression that he is viewed as a returned war bore. Elizabeth and Fanny had symbolised civilisation, life and hope to George. But the war had created an abyss between them. The ladies see a vast degeneration in George.

'It's quite useless,' said Elizabeth; 'he's done for. He'll never be able to recover. So we may as well accept it. What was rare and beautiful in him is as much dead now as if he were lying under the ground in France.' And Fanny agreed.[35]

Thus the story of George Winterbourne's boredom ends with boredom as a reaction to the sacrifice of a nation's individuals in the slaughter and horror of the trenches. Thus, war first attracts boys away from civil boredom. Once they're in the war, war bores and degenerates them. Then, if and when soldiers return from the war zones, they bore the ones who sent them to war. The loop of boredom is complete.

We will now turn to another case study of boredom, albeit quite at the other extreme of the military boredom scale. Though less desperate and degenerating and seemingly absent, boredom is still there—in the *anticipation* of it.

3. Norwegian Infantry Soldiers in Afghanistan

In a case study of 15 Norwegian soldiers serving in Afghanistan, mainly reconnaissance troops, I tried out a method of offering these soldiers a diary in which they should comment on 25 (moderately provocative) statements (one each day), carefully selected to cover feelings and mental states associated with boredom.[36] This took place about a month into their tour.

Regarding their service, the overall material seems to express a general confidence in and happiness about their deployment to Afghanistan. When explicitly asked to give reasons for why they wished to embark on such a service, they highlighted the following motivational 'clusters': excitement/challenge; adventure/experience; earn some good money; to do something together with the military team; to see another culture. This fits neatly with the so-called 'post-modern soldier' as described by military sociologists such as Fabrizio Battistelli and Charles Moskos.[37] They were also asked whether their expectations had been met so far into their deployment. The reaction was almost unanimous: 8 replied that 'all expectations' had been met, whereas 6 said 'largely met,' one 'partly met,' and none 'not met.' This reinforces the general picture of a group of soldiers that was very confident overall. However, this does not mean that boredom was wholly absent as an issue, whether in the form of an actual challenge to their confidence or as a potential threat not yet encountered.

For most soldiers, their daily tasks, especially going out of the camp on patrols, but also the opportunities they had to help Afghan people, were perceived as highly meaningful. Nevertheless, one may contend that boredom was still present among many of these troops, simply by manifesting itself in the form of its absence, which made it consciously present. This could be seen in rather elaborate strategies for diversion and the avoidance of boredom. It seems therefore that many of the soldiers perceived boredom as a potential which was there, lurking around them. Consequently, in order to prevent it from 'seizing' them, they had to 'fill time' with various kinds of diversion. In most cases, boredom was successfully prevented, though it was still there in their consciousness.

The greatest risk to them in their handling of 'potential boredom,' however, was the risk of exhaustion. To be alert for long periods, to tip-toe on every patrol, to be too excited too often, was experienced as quite meaningful, but it would also lead to a deterioration of their ability to stay alert and fully attentive after a while. Hence, they needed rest, which again would provide a vitalising restlessness (boredom). This issue surfaced as the soldiers were asked to comment on the following statement provided by the researchers for their diaries:

> It is good for human beings to be bored.

This apparent quotation is not actually a real quotation. Even though the gnomic style may look a bit like a quotation from a famous philosopher or novelist, it was just one of the statements included in the list for use in connection with the soldiers' diaries. It may also look like a common saying, reminding people of childhood or school. One of the soldiers treats it in this way, saying that 'this is a well-known statement.' Another soldier smells the deception and commences his answer thus: 'Hmmm...This really sounds like something from a professor.'

The reactions can be divided into two camps. The minority camp could be defined by the reaction of a soldier, imitating the same gnomic style: 'People who are bored are not capable of seeing the possibilities,' he says, or in the more resolute fashion of another soldier: 'Bullshit! If so, one shouldn't send them to Afghanistan....' The understanding of boredom is that it does not need to occur, given that one takes the initiative to fill one's spare time or free time before boredom sets in.

People in this camp tend to see boredom as something negative and threatening. The opinion of one soldier is that it is not good for people to be bored:

> Time passes slowly and things get boring. I feel that something should be happening the whole time. Otherwise, I don't have much to say on this matter. I'm afraid the question is too complex for me to answer. I'm just a simple grunt...

The problem that quickly emerges is an over-filling of time. It is as if many of them hunger after less filled time. Some of them have been working every single day for the last month or more and have been looking forward to a day off the next Sunday (as they wrote in reaction to this statement). But the officer may have activities scheduled for this Sunday as well. Desperation draws nearer and they long for some boredom, says one soldier.

The other camp is characterised by an affinity for affirming the 'quote' above (the statement on which they were asked to comment), even though they may find it problematic to have too much time for leisure or reflection. They find boredom necessary for reflection, providing room for afterthought and imagination. One soldier eloquently describes this room:

> In principle, being bored is a real drag. At the same time, you are challenged to do something about it. I've been on a lot of skiing expeditions in my life and it can be pretty boring to look into the white 'nothingness.' After a week you have really finished pondering things you normally don't get time to think over. My way of solving this problem is to use my imagination. I put myself in a fictive story and form my future myself, using myself as both the main character and the hero of the story. Other times I plan new trips or projects I wish to start on when I get back home. ... So, based on the fact that I'm both more effective and more constructive when I'm bored, I agree with the statement [that it is good for people to be bored].

We see how boredom not only creates room for reflection and imagination, but also a narrative room, which is important if you want to 'write' a comprehensive, coherent meaning into the story of your life. However, this soldier's experience from skiing expeditions does not necessarily apply to his service in Afghanistan: 'To sit here and feel bored rarely provides me with constructive energy. I think the context of boredom is decisive for how you spend your time.'

Another soldier, who also belongs to the camp that sees boredom as constructive, also uses words such as 'imagination.' Moreover, he also puts boredom into a particular context, namely that of deprivation. He sees boredom as a kind of deprivation, which is hardly the experience of most people he knows. He maintains that these people only need to 'act,' not to 'think.' The consequence is that they hardly appreciate anything.

> If you are bored ... for a period, I think you would appreciate normal things in another way. That's my experience at least. But one should keep this in mind as days become normal again.

This view is implicit in the reaction of a third soldier:

> People benefit, if not from boredom, then at least from having some time alone with no other demands than just being able to think. Some people would find this boring, whereas I find it perfectly fine. When you are far away from Norway and the things you are familiar with, you learn to understand yourself, in particular the way you learn about what is really important to you. And that can't be bad for anybody.

One could say that for this soldier boredom has the function of deepening his understanding of himself, as well as providing a clearer understanding of which values count the most to him. A fourth soldier who emphasises the constructive side of boredom, says that 'nobody enjoys being bored.' Yet, he still sees the positive potential inherent in boredom. Hence, he embraces the opinion that one probably benefits from being bored: 'You get time to think and reflect about various issues and things, perhaps something you would like to do when you get home.' In his case, this obviously also applies to his situation as a soldier in Afghanistan. Time is something one 'gets,' time is a *gift*, not an enemy. However, as a 'borderline' representative, i.e., with an affinity for both the constructive and destructive elements attached to the state of boredom, he also uses rather strong negative language about boredom: 'I hate being bored.' It seems that boredom is a rather ambiguous state of mind, being both a friend and a foe.

One way of keeping the constructive and destructive aspects of boredom together, is to understand boredom, as perceived by the soldiers, as *a threshold phenomenon.* If understood thus, boredom is something that occurs at a certain point and at a certain level. Another soldier, one of the ones who uses the word 'imagination' in connection with boredom, clearly depicts it as a threshold phenomenon:

> As a child, I always heard that it was healthy to be a little bit bored; good for the imagination. Yet, to be very bored is not good. You can become restless and do stupid things that are not very sensible in a conflict zone.

Consequently, one should keep an eye on this threshold.

It is my suggestion that the threshold perspective is indirectly reflected in the responses of the soldiers to the various statements to which I asked them to respond. It is a bad thing in itself to be bored a lot and it is particularly bad in a conflict zone. Soldiers may start doing things that are counterproductive to their mission. This astutely sums up the whole problem of military boredom. This is not just a human phenomenon observed in the armed forces. It is also specifically related to the success of the mission to which one is attached; if boredom succeeds, the mission will fail.

4. Interpretation and Conclusion

We have so far described the complexity of military boredom, both through the experiences of George Winterbourne from the First World War and Norwegian soldiers operating in a rather quiet situation in the Afghan mountains (which some months later turned quite violent). In my conclusion, which relates to primarily the latter case study, I would like to say a few words about boredom and meaning. This may in particular have relevance for the development of adequate military leadership in situations where either manifest or anticipated boredom marks the states of minds of the soldiers.

We have already seen how the soldiers report a high degree of confidence and meaning in their assignments and general tasks. In fact, the degree of *meaning* (or the lack of it) is the factor that determines whether something is experienced as boring or not. To give just one example, one may have a great deal to do or very little to do, but, as long as this is experienced as meaningful, one is not bored. Consequently, both a busy businessman and a retired businessman with almost nothing to do may be equally happy, provided that their days are perceived as equally meaningful.

Mihaly Csikszentmihalyi's theory of *flow* may be useful for our understanding of boredom among the Norwegian soldiers. Csikszentmihalyi calls what makes us happy 'flow,' or what he calls 'optimal experience.' Flow can be seen as a finely-tuned course between boredom and anxiety and can only take place when challenges are well-matched to our skills and experience in the sense that challenges should be slightly more demanding than merely requiring and using one's existing skills and knowledge. If not, challenges will cease to be challenges and will sooner or later develop into a state of boredom. On the other hand, however, too demanding tasks may incapacitate through the laming impact of anxiety.

The diaries of the Norwegian soldiers seem to express a high degree of flow. Almost none of the soldiers would have preferred to be doing something else, or be somewhere else and most of them seem to enjoy the assignments and the fellowship of comrades. But it still seems to be a vulnerable flow, prone to a certain kind of boredom.

Dissatisfaction is mentioned in *Flow* and it can be seen as the opposite of a good and meaningful life. According to Csikszentmihalyi, *dissatisfaction* can consist of one of the following three elements: (a) lack of variety and challenge, (b) conflicts, (c) burnout.[38] Viewed in light of these options, our material (based on the entire collection of diaries) shows that (a) service in Afghanistan displays a rather high degree of variation and challenges, (b) that hardly any conflicts are mentioned and (c) burnout may be a problem, based on our demonstration of exhaustion above. Hence, the level of satisfaction is rather high and, if there is a potential threat to it, it mainly has to do with exhaustion and too few breaks.

Similarly, Csikszentmihalyi mentions in his book what constitutes meaning, namely three elements: (a) purpose; i.e., ultimate goal, (b) resolution; intention and (c) harmony, i.e., cohesion and order among otherwise fragmented and unconnected events or elements.[39] In short, his approach is this simple: 'the meaning of life is meaning.' Let us just see how the two former criteria within this formula may apply in the case of Norwegian soldiers in Afghanistan.

Firstly, based on our statement about the perceived morality of the ISAF operation, the ultimate purpose of their deployment seems to have been fulfilled. Even the micro-goals of their service appear to have a point and to make sense. In our study, it is only one soldier, who is the driver for an officer, who does not see the point of what he is doing. Others may express similar feelings in passing, but this is not a prominent feature of their diaries.

Secondly, resolution has already been identified in what was said about their will to fill their time. These soldiers do not sit still and wait for orders. This may reflect the Norwegian military leadership philosophy, which is very decentralised and puts a large responsibility for judgements and decisions on the shoulders not only of officers on all levels, but also of privates. We noted, however, that the will to fill time contains an inherent danger of overfilling. Paradoxically, this can be understood as having to do with what they complained about in their officers, namely that there was too little time left for breaks and rest. But it can also be understood as a contribution to rest and recreation, i.e., by sunbathing, by watching movies, by surfing on the internet, they get a necessary break (di-version!) from their conventional modes of operating. The test seems to be whether the means that are used to fill the downtime are meaningful in themselves, as in this quote from one of the officers:

> I do not usually get bored. I have the ability to fill 'spare time' with meaningful activities, or deliberately abstain from doing anything, which is also meaningful. To be able to relax is an important skill.

Expressions like this should be analysed and interpreted carefully if one is to acquire an adequate understanding of military boredom in current operations, both as an explicit and implicit phenomenon. And it seems that the perception of boredom as either hostile/destructive or as something friendly/constructive may be a useful pair of concepts for analysis.

Notes

[1] Anthony Swafford is both the main character and the author of this book, which has also been made a film (directed by S. Mendes), covering a rather bleak but intense experience of boredom within an environment of waiting and lack of use of competence (both main characters are snipers) and hence a deep sense of meaninglessness.
[2] B. Edge & A. Cup, 'X-Treme Gambling, Part 1. Russian Roulette: The Game You Can't Win,' *GamblingGates.com*, Viewed on 21 March 2005, <http://www.gamblinggates.com/Print/Russian-Roulette8328.html>.
[3] D.R. Segal & J. Harris, 'Observations from the Sinai: The Boredom Factor,' *Armed Forces and Society*, 1985, p. 235.
[4] Department of the Army, 'Peacekeeping,' *FM 7-98 OPERATIONS IN A LOW-INTENSITY CONFLICT*, 19 October 1992, Viewed on 29 March 2005, <http://www.globalsecurity.org/military/library/policy/army/fm/7-EF73h2>.
[5] For example cf. Gift Baskets Royale, 'Specialty and Military Care Packages,' Viewed on 19 April 2008, <http://www.giftbasketsroyale.com/page/928797>.
[6] S.D. Healy, *Boredom, Self and Culture*, Fairleigh Dickenson University Press, Rutherford/Madison/Teaneck, 1984, p. 28.
[7] P.T. Bartone, 'Vulnerability, Trauma and War: Toward Understanding Abu Ghraib Iraqi Prisoner Abuse,' *Who Gets PTSD? Issues of Posttraumatic Stress Vulnerability*, J.M. Violanti & D. Paton (ed.), Charles C. Thomas, Springfield, Ill.
[8] Both cases appear in my book *Enduring Military Boredom: 1750 to the Present*, co-authored with P.O. Brunstad, Palgrave Macmillan, Basingstoke, 2009.

[9] Quoted by C. Ridgeway in his introduction to the 1984 edition of *Death of a Hero* by R. Aldington, The Hogarth Press, London, [1929]1984.
[10] Ibid., p. 264.
[11] Ibid., p. 288.
[12] J. Bourke, *An Intimate History of Killing: Face-to-Face Killing in Twentieth Century Warfare*, Granta Books, London, 2000, pp. 306f.
[13] Aldington, op. cit., pp. 241, 250, & 261.
[14] Ibid., p. 238.
[15] Ibid., p. 242.
[16] Ibid., p. 251.
[17] Ibid., p. 264.
[18] Ibid., p. 266f.
[19] Ibid., p. 280.
[20] Ibid., p. 267.
[21] Ibid., p. 250.
[22] Ibid., p. 242.
[23] Ibid., p. 253.
[24] Ibid.
[25] Ibid., p. 258.
[26] Ibid., p. 227.
[27] Ibid., p. 257.
[28] Ibid., p. 289.
[29] Ibid., p. 264.
[30] Quoted in P Fussell, *The Great War and Modern Memory,* Oxford University Press, Oxford, 2000, p. 149.
[31] Aldington, op. cit., p. 287.
[32] Ibid., pp. 296 and 307.
[33] Ibid., p. 323.
[34] Ibid., p. 346.
[35] Ibid., p. 228.
[36] The range of themes included: attention and awareness vs. drowsiness and lethargy; explicit experience and interpretation of boredom; friendship and fellowship vs. solitude and loneliness; meaningfulness vs. lack of meaning (and morality); joy and happiness vs. sadness; variation, excitement and action vs. monotony; the importance of individual moral qualities; coping strategies; means of keeping in touch with 'those back home' (letters, emails etc.); unrest, restlessness; boredom as being 'too much;' motivations and expectations; caring (for) vs. ignorance and negligence; possible destructive emotions; who is responsible for boredom.
[37] Cf. FI Battistelli, 'Peacekeeping and the Postmodern Soldier,' *Armed Forces & Society*, Vol. 23, Iss. 3, 1997, pp. 467-84; C.C. Moskos, 'Toward a Postmodern Military: The United States as a Paradigm,' *The Postmodern*

Military: Armed Forces after the Cold War, C.C. Moskos, et al. (eds), Oxford University Press, Oxford, pp. 14-31.
[38] M. Csikszentmihalyi, *Flow: The Classic Work on How to Achieve Happiness*, Rider, London, 2002, Chapter 10 ('The Making of Meaning'), in particular p. 160.
[39] Ibid., p. 216.

Bibliography

Aldington, R., *Death of a Hero*. The Hogarth Press, London, [1929]1984.

Bartone, P.T., 'Vulnerability, Trauma and War: Toward Understanding Abu Ghraib Iraqi Prisoner Abuse.' *Who Gets PTSD? Issues of Posttraumatic Stress. Vulnerability*. Violanti, J.M. & Paton, D. (eds), Charles C. Thomas, Springfield, IL, 2006.

Battistelli, F., 'Peacekeeping and the Postmodern Soldier.' *Armed Forces & Society*. Vol. 23, Iss. 3, 1997, pp. 467-84.

Bourke, J., *An Intimate History of Killing: Face-to-Face Killing in Twentieth Century Warfare*. Granta Books, London, 2000.

Csikszentmihalyi, M., *Flow: The Classic Work on How to Achieve Happiness*. Rider, London, 2002.

Department of the Army, 'Peacekeeping.' *FM 7-98 OPERATIONS IN A LOW-INTENSITY CONFLICT*. 19 October 1992, Viewed on 29 March 2005, <http://www.globalsecurity.org/military/library/policy/army/fm/7-EF73h2>.

Edge, B. & Cup, A., 'X-Treme Gambling, Part 1, Russian Roulette: The Game You Can't Win.' *GamblingGates.com*. Undated, Viewed on 21 March 2005, <http://www.gamblinggates.com/Print/Russian-Roulette8328.html>.

Fussell, P., *The Great War and Modern Memory*. Oxford University Press, Oxford, 2000.

Healy, S.D., *Boredom, Self and Culture*. Fairleigh Dickenson University Press, Rutherford/Madison/Teaneck, 1984.

Kierkegaard, S., *Enten-Eller [Either-Or], 3rd Edition, Vol. 2*. Gyldendal, Copenhagen, 1982.

Moskos, C.C., 'Toward a Postmodern Military: The United States as a Paradigm.' *The Postmodern Military: Armed Forces after the Cold War*. C. Moskos, C. et al. (eds), Oxford University Press, Oxford, 2000.

Mæland, B. and Brunstad, P.O., *Enduring Military Boredom: 1750 to the Present.* PalgraveMacmillan, Basingstoke, Houndmills, 2009.

Segal, D.R. and Harris, J., 'Observations from the Sinai: The Boredom Factor.' *Armed Forces and Society*. Vol. 11, 1985, pp. 235-248.

Swafford, A., *Jarhead: A Soldier's Story of Modern War.* Scribner, London, 2003.

Bård Mæland is a professor of systematic theology at the School of Mission and Theology, Stavanger, Norway where he now serves as the Rector. He used to serve as a Chaplain of the Norwegian Defence and is also the founding editor of *Journal of Military Ethics (informaworld.com/jmile).*

www.ingramcontent.com/pod-product-compliance
Lightning Source LLC
LaVergne TN
LVHW102227300926
849160LV00005B/48